Handbook of Mariculture

Handbook of Mariculture

Edited by:

D. M. Pawar

Delve
PUBLISHING

www.delvepublishing.com

Handbook of Mariculture

D. M. Pawar

Delve Publishing
224 Shoreacres Road
Burlington, ON L7L 2H2
Canada
www.delvepublishing.com
Email: orders@arclereducation.com

ISBN: 978-1-77469-131-1 (Hardcover)

ABOUT THE EDITOR

Dr. Dnyaneshwar M. Pawar (1985) is presently serving as Assistant Professor, Department of Plant Pathology, college of Agriculture, Naigaon. He obtained his B. Sc. (Ag.) in 2008 from Marathwada Agriculture University, Parbhani (Maharashtra). M. Sc. (Ag.) 2010 and Ph. D. (2014) in the discipline of Plant Pathology from Navsari Agricultural University, Navsari (Gujarat). He started his career as Assistant professor in 2014. His field of specialization is Plant Pathology (mycology and virology). He has published 01 book and 11 research paper in national and international journals of reputed. He has participated many National, State Seminar and Symposiums. He has written 02 practical manuals for under graduate course for the benefit of students besides several Radio talk, Lectures in farmer training programme. Dr. Dnyaneshwar has also written 3 popular articles in reputed Magazines. He has excellence teacher award on 2019 from academy for Environment and Life Science.

TABLE OF CONTENTS

LIST OF FIGURES

LIST OF TABLES

LIST OF ABBREVIATIONS

ADM	Archer Daniels Midland
ASC	Aquaculture Stewardship Council
BIFS	Brackish Water Integrated Farming System
BMU	Beach Management Units
CCPs	Community Fishing Councils
CCRF	Code of Conduct for Responsible Fisheries
COMESA	Common Market for Eastern and Southern Africa
EAA	Ecosystem Approach to Aquaculture
EAF	Ecosystem Approach to Fisheries
EEZ	Exclusive Economic Zone
EIA	Environmental Impact Assessment
EMS	Environmental Management System
FAO	Food and Agriculture Organization
GAA	Global Aquaculture Alliance
GDP	Gross Domestic Product
GESAMP	Joint Group of Experts on the Scientific Aspects of Marine Pollution
IFOAM	International Federation of Organic Agriculture Movements Basic Standards
IMTA	Integrated Multi-Trophic Aquaculture
INFOPECHE	Intergovernmental Organization for Marketing Information and Cooperation Services for Fishery Products in Africa
ISO	International Standards Organization
IUU	Illegal, Unreported, and Unregulated
MCE	Multicriteria Evaluation
MSI	Mariculture Sustainability Index
NOAA	National Oceanic and Atmospheric Administration
PAS	Partitioned Aquaculture System
RAS	Recirculating Aquaculture Systems
SCBA	Social Costs Benefit Analysis

UV	Ultraviolet
WHO	World Health Organization
$ZnSO_4$	Zinc Sulfate

GLOSSARY

A

Amberjack – is an Atlantic and Pacific fish in the genus Seriola of the family Carangidae.

Aquaculture – also known as aquafarming, is the farming of fish, crustaceans, mollusks, aquatic plants, algae, and other organisms. The aquatic counterpart of agriculture, has grown rapidly in recent decades, and today it produces almost as much fish and shellfish as fisheries. Aquaculture is the main means for obtaining more food from our aquatic environments in the future.

Articulated – having two or more sections connected by a flexible joint.

B

Bio Deposition – deposition of biogenic material on the bottom of a body of water, usually in the forms of feces and pseudofeces.

Bio Mitigation – steps taken to avoid or minimize negative environmental impacts.

Biogeochemical – in ecology and Earth science, a biogeochemical cycle is a pathway by which a chemical substance is turned over or moves through the biotic and the abiotic compartments of Earth.

Biomass – the total quantity or weight of organisms in a given area or volume.

Bivalves – as a group have no head and they lack some usual molluscan organs like the radula and the odontophore.

Brackish Water – is a broad term used to describe water that is more saline than freshwater but less saline than true marine environments.

C

Coastal – these areas are local administrative units (LAUs) that are bordering or close to a coastline.

Contaminants – a polluting or poisonous substance that makes something impure.

Cultivation – the act of preparing land and growing crops on it, or the act of growing a particular crop.

E

Ecology – is the study of the relationships between living organisms, including humans, and their physical environment.

Economics – is the social science that studies how people interact with value; in particular, the production, distribution, and consumption of goods and services.

Estuaries – an estuary is a partially enclosed coastal body of brackish water with one or more rivers or streams flowing into it, and with a free connection to the open sea.

F

Fabricated – invent (something) in order to deceive.

Farming – it is growing crops or keeping animals by people for food and raw materials. People may have started farming because the weather and soil began to change. Farming can feed many more people than hunter-gatherers can feed on the same amount of land.

Fish Farms – a place where fish are bred for commercial purposes.

Fishermen – a person who catches fish either as a job or as a sport.

Flocculation – in the field of chemistry, is a process by which colloidal particles come out of suspension to sediment under the form of floc or flake, either spontaneously or due to the addition of a clarifying agent.

H

Habitats – in ecology, the term habitat summarizes the array of resources, physical, and biotic factors that are present in an area, such as to support the survival and reproduction of a particular species.

Harvesting – is the process of gathering a ripe crop from the fields.

Hatcheries – it is a facility where eggs are hatched under artificial conditions, especially those of fish, poultry or even turtles.

Herbivorous – is an animal anatomically and physiologically adapted to eating plant material, for example foliage or marine algae, for the main component of its diet.

I

Imitate – take or follow as a model.

J

Juxtaposition – is an act or instance of placing two elements close together or side by side.

M

Macroalgae – is a collective term used for seaweeds and other benthic (attached to the bottom) marine algae that are generally visible to the naked eye.

Mangrove Fringes – fringe mangrove forests are the typical and classic mangrove zonation pattern.

Manpower – the people that you need to do a particular job.

Mariculture – is a specialized branch of aquaculture involving the cultivation of marine organisms for food and other products in the open ocean of offshore aquaculture, an enclosed section of the ocean, or in tanks, ponds or raceways which are filled with seawater.

Microalgae – are unicellular photosynthetic micro-organisms, living in saline or freshwater environments, that convert sunlight, water, and carbon dioxide to algal biomass.

Mullets – any of various chiefly marine fish that are widely caught for food.
Multicellular – these organisms are organisms that consist of more than one cell, in contrast to unicellular organisms.

O

Optimizing – an act, process, or methodology of making something (such as a design, system, or decision) as fully perfect, functional, or effective as possible specifically, the mathematical procedures (such as finding the maximum of a function) involved in this.
Oysters – is the common name for a number of different families of salt-water bivalve mollusks that live in marine or brackish habitats.

P

Parasites – is an organism that lives on or in a host organism and gets its food from or at the expense of its host.
Pathogenic – (of a bacterium, virus, or other microorganism) causing disease.
Penaeids – is a family of marine crustaceans in the suborder Dendrobranchiata, which are often referred to as penaeid shrimp or penaeid prawns.
Phytoplankton – are the autotrophic components of the plankton community and a key part of ocean and freshwater ecosystems.
Piscivore – a piscivore is a carnivorous animal that eats primarily fish.
Polychaetes – also known as the bristle worms or polychaetes, are a paraphyletic class of annelid worms, generally marine.
Polyculture – is a form of agriculture in which more than one species is grown at the same time and place in imitation of the diversity of natural ecosystems.
Pseudofeces – or pseudofeces are a specialized method of expulsion that filter-feeding bivalve mollusks use in order to get rid of suspended particles such as particles of grit which cannot be used as food, and which have been rejected by the animal.

S

Sea Urchins – are typically spiny, globular animals, echinoderms in the class Echinoidea.
Seaweeds – is the common name for countless species of marine plants and algae that grow in the ocean as well as in rivers, lakes, and other water bodies.
Shrimp – a decapod crustacean with an elongated body and a primarily swimming mode of locomotion—most commonly Caridea and Dendrobranchiata, typically marine and frequently of commercial importance as food.
Socio-Economic – is the social science that studies how economic activity affects and is shaped by social processes.
Sterilization – refers to any process that removes, kills, or deactivates all forms of life and other biological agents like prions present in a specific surface, object, or fluid, for example food or biological culture media.
Sustainable – meeting our own needs without compromising the ability of future generations to meet their own needs. In addition to natural resources, we also need social and economic resources.

Sustenance – the food and drink that people, animals, and plants need in order to live and stay healthy.

T

Threadfin – are silvery grey perciform fish of the family Polynemidae, found in tropical to subtropical waters throughout the world, the threadfin family contains eight genera and about 40 species.

Threshold – a strip of wood or stone forming the bottom of a doorway and crossed in entering a house or room.

Tilapia – an African freshwater cichlid fish that has been widely introduced to many areas for food, such as St. Peter's fish. This is the common name for nearly a 100 species of cichlid fish from the coelotilapine, coptodonine, heterotilapine, oreochromine, pelmatolapiine, and tilapiine tribes, with the economically most important species placed in the Coptodonini and Oreochromini.

V

Vulnerability – the quality or state of being exposed to the possibility of being attacked or harmed, either physically or emotionally.

PREFACE

This book takes the readers through several different stages of mariculture and its types, tropical mariculture, environmental impacts of mariculture. It also gives a brief overview of mariculture, economic, and social impacts. This book sheds light on mariculture and farming technology, methodologies related to mariculture, and its sustainability and regulations, along with explaining the future of aquaculture and mariculture.

The first chapter stresses on the basic overview of mariculture, trends in aquaculture, the different types of mariculture. It focuses on the sea cage farming method, longline farming as well as raceway farming. This chapter will also emphasize on the structures of mariculture operations along with the marine environment and structure for cultures.

The second chapter takes the readers through the concept of tropical mariculture, fish farming in the tropics, ocean farming in the tropics, and sub-tropics. This chapter will provide highlights on the integrated multi-trophic aquaculture that is the IMTA, its concepts, and current systems. It also states the integrated tropical mariculture.

Then, the third chapter explains the environmental impacts of mariculture, ecological effects of bivalve mariculture, environmental impacts of shellfish aquaculture. It also provides an insight into the different impacts of mariculture on water quality. It also explains the impacts of mariculture on sediment.

The fourth chapter introduces the readers about the mariculture, economic, and social impacts, the socio-economic aspects, contribution to livelihoods, role in social structure. This chapter also explains the social risk in mariculture, the economic impacts of provincial mariculture, along with the socio-economic impacts of mariculture.

The fifth chapter throws light on mariculture and farming technologies. Existing major mariculture species and farming, as well as grow-out technology. It further explains the eight digital technologies disrupting aquaculture. This chapter also contains the integrated fish farming, backyard re-circulatory aquaculture system, and the different types of farming in mariculture.

The sixth chapter takes the readers through concepts and methodologies related to mariculture, the identification of the economic, social, and environmental costs and benefits of aquaculture. The readers are then told about the aquaculture methods and practices and the recirculating aquaculture system.

The seventh chapter explains the sustainability and regulations involved in mariculture, aquaculture regulatory frameworks, along with illustrating the mariculture guidelines related to biodiversity. This chapter also emphasizes the development of sustainable methods such as lagoons and artificial wetlands, sustainable feed management, and the use of chemicals.

The last chapter of this book sheds light on the future of mariculture as well as aquaculture, sustainable growth, and development in various countries, along with defining the diseases. This chapter also mentions the sea change that is the new horizons in aquaculture, new options from start-ups. It sheds light on the feeding of farmed fish and the integrated mariculture and its role in the future development of aquaculture.

This book has been designed to suit the knowledge and pursuit of the researcher and scholars and to empower them with the concepts of mariculture and aquaculture along with defining the various forms and aspects of the mariculture, its types, and trends as well as its development in the future so that they are updated with the information.

I hope that the readers find the book explanatory and insightful and that this book is referred by scholars across various fields.

Chapter 1

Introduction to Mariculture

CONTENTS

This chapter provides an introduction to mariculture. It begins with highlighting the trends in aquaculture, along with the various types of mariculture that are seawater ponds, tank farming, algaculture, sea cage farming, longline farming, raceway farming, fish hatcheries, integrated multitrophic aquaculture.

It also addresses the structures for mariculture operations, such as the requirements for coastal mariculture development in earth ponds, culture methods in earth ponds, and environmental considerations. In the end, it sheds light on the marine environment and structures for culture.

1.1. INTRODUCTION

Aquaculture is prominent and is well known from other aquatic production, for example, fishing by the extent of the interference of human beings and regulation that is possible and attainable as stated by Anderson in 2002. The methods of production in aquaculture are demonstrated by various factors such as the biological, technological, economic, and environmental aspects. Although, the prime factor is that various aspects that are carried out in the procedure of the production be conducted under the management and control of humans.

Through this control, innovation and restructure are made possible, and in addition to this, it is also significant for the technological development that happens rapidly, which has been responsible for the growth and development of production, since the early 1970s. As stated by the United Nations Food and Agriculture Organization (FAO), aquaculture refers to the "farming of aquatic organisms such as fish, mollusks, crustaceans along with the aquatic plants."

Farming suggests the interposition in the process of rearing in order to enhance production, such as stocking, feeding, security measures, etc. The beginning of aquaculture has its roots from millennia; however, its exact origins are unspecified and unknown, and a large number of organisms that humans are dependent on for acquiring protein and nourishment that comes from the sea.

In recent years, around 16% of the animal protein which is absorbed and utilized by the world's population is drawn from fish, and across 1 billion people all over the globe rely on fish as their major source of animal protein.

> During the 1970s, the year which is often labelled as the "blue revolution" established as humans racked up knowledge of aquaculture allowed for the foundation of the farming practices which includes two types i.e., semi-intensive and intensive.

As a result, producers were in a position of influencing the growing conditions of the fish by the medium of feeding, breeding, and so on, and the production pattern was shut down for an enlarging number of species. The growing control of the process of production delegated numerous productivity-enhancing innovations and changes to occur.

The better productivity resulted in a reduction in the costs of production, and with a specified price, this gave rise to more profitable and cost-effective production. In almost every part of the world various number of species are being farmed, in freshwater as well as in saltwater.

Additionally, a number of distinct methods of production are being implemented, adapted to individual species, environments, and economic conditions. These techniques inculcate ponds, pens, raceways, ropes, cages, tanks, and the systems of closed circulations.

1.2. TRENDS IN AQUACULTURE

Since the growth potential for wild fisheries is restricted, it is extensive for aquaculture. Aquaculture is a production technology having its emergence in Egypt and China many centuries ago. Beginning in the 1970s, an essential transition took place as better regulation and hold over the processes of production which resulted in developing and advancing a number of new technologies and various practices of production.

These transformations are adequately enhanced and improved the competitiveness of aquaculture products both as sources of basic food as well as the cash crops. The joined forces and effect of productivity and the growth in market has made aquaculture the world's fastest expanding animal-based food sector of the last generations and centuries (OECD, 2010).

The species that are constructed and produced in the process of aquaculture is nearly as large as the species present in wild fisheries. The production of aquaculture includes kelp (seaweed), mussels, crustaceans, carps, tilapia, salmon, seabass, shrimp, etc. (Figure 1.1).

Figure 1.1. An illustration of seaweed.

Source: Image by Pxhere.

While aquaculture has been a success as it gives favorable outcomes in the context of increased production, it also encounters strong opposition and conflict in various countries because the new technologies that are enabling the increased aquaculture production are having adverse effects on the environment. The use of these technologies is depriving the world from its natural resources.

There are numerous examples of unsustainable as well as sustainable aquaculture practices. It is of the highest importance to encourage sustainable practices and discourage ruining of locations and causing negative impacts on the environment.

1.3. TYPES OF MARICULTURE

A very brief summary of the important types of mariculture is defined in subsections.

1.3.1. Seawater Ponds

In seawater ponds, aquatic species are grown in ponds which acquires water from the sea. This enjoys the advantage that plankton which are present in the seawater can be made use of as a source of food (e.g., prawns; fish; eels; crayfish).

1.3.2. Tank Farming

Some species are used to growing well in tanks which are oxygenated and have a constant exchange of water so as to retain the diffused oxygen levels high and it also extracts wastes (e.g., prawn brood stock tanks; prawn culture tanks; barramundi).

1.3.3. Algaculture

Algaculture is defined as the farming of different species of algae. But in order to begin with the basics one needs to ask questions like what even is algae? So as to answer this, algae is defined as "a plant or plantlike organism of any of several phyla, divisions, or classes of chiefly aquatic usually chlorophyll-containing nonvascular organisms of polyphyletic origin that most commonly include the green, yellow-green, brown, and red algae in the eukaryotes and especially formerly the cyanobacteria in the prokaryotes" (Merriam-Webster Dictionary). Although the definition stated above is complex in nature which sometimes leads to confusion. But the truth is, there is a plenty of indistinct and blurry terms used to elaborate algae, and that is probably because the taxonomical classifications of algae are dynamic in nature, which keeps on evolving (Figure 1.2).

Figure 1.2. An illustration of rhodophytes.

Source: Image by Flickr.

Primarily to encapsulate this, algae is identical to a plant, but is not truly a plant. To justify this, all plants belong to the Kingdom Planate, whereas all algae does not. Moreover, plants are multicellular, vascular, and are found mostly on land.

On the other hand, algae are multi or unicellular, non-vascular, and are mostly found in the sea, submerged or either floating. It also comes under the various Kingdoms. The chlorophytes and rhodophytes that is the green and red algae belongs to the Kingdom Plantae, the Kingdom Chromista includes phaeophytes that is the brown algae, the Euglenophytes such as the Euglena resides in the Kingdom Protista, while the cyanobacteria that is the blue green algae can be found in the Kingdom Bacteria.

Although plants and algae contain two major characteristics that are common to each other; Firstly, they are both eukaryotes, and secondly, they are both photosynthetic and thus contain chlorophyll or chloroplasts.

In order to define algae in a more elaborated and broad terms, the two major types of algae that can be found are: Macroalgae and Microalgae. Macroalgae is large aquatic photosynthetic plants that are visible without the help of a microscope and are often termed as multicellular. Kelp and seaweeds are the most common example of macroalgae.

Whereas, microalgae or in other words microphytes are microscopic algae that are not seen by the naked eye, and are often referred to as unicellular, or present in chains and groups. Phytoplankton is the common example of microalgae which are usually found in freshwater and marine systems.

Several of the most prominent species that are used in Algaculture are: Seaweed (Nori, etc.), Spirulina, Chlorella, Kelp, Chondrus Crispus aka 'Irish sea moss,' as well as most of the red alga (Rhodophyta) as they account for producing carrageenan.

But how are algae farmed? To grow the alga aquaculture techniques are used. It is produced in ponds or tanks by adding nutrients and carbon dioxide to the system. Once harvested, the alga must be dried. There are three major methods to accomplish this by: Flocculation, Centrifugation, and Micro-screening.

1. **Flocculation:** It takes place when solids transpire from the suspension (in a floc/flake) all of a sudden or because of a clarifying agent which is added. The biggest drawback of this process is that it is way too expensive and thus only large-scale

industries make use of it as small-scale industries are not able to meet the costs incurred on it (Figure 1.3).

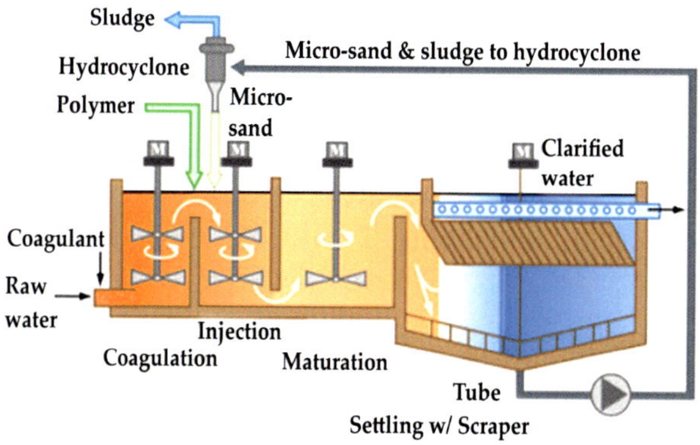

Figure 1.3. An illustration of ballasted flocculation.

Source: Image by Flickr.

2. **Centrifugation:** It is a process that is responsible for detaching algae from water. The machine spins and forces the algae up against the sides, constructing a paste. The water flies to the top as it spins instead of staying with the algae against the sides. This method does not require a large amount of money, it is helpful for industries who are capable of making medium expenses.
3. **Micro Screening:** It is a process of membrane separation that is done by means of filtration system. It puts to use the properties of water-cohesion and tension-in order to lead to the Capillary effect, which eventually acts as a catalyst for water to "climb up" or in other words work against gravity and separate out from the alga. After this, a fine mesh screen is employed to sift algae out.

After carrying out the above-mentioned methods, these algae once it is abandoned and dried, it can serve to unbelievable number of purposes and multiple uses. Some of the uses of algae can be, it is in the food people eat, colorants/dyes, fertilizer, beauty products, pharmaceuticals, traditional medicine, and many more.

For instance, spirulina, and chlorella particularly are termed as "Superfoods." Carrageenan and Agar are withdrawn and can be utilized in many food items in a form of stabilizer or thickener. In the beauty industry algae holds a pivotal position as a miracle worker due to its vitamins, minerals, antioxidants, cell renewal (anti-aging) ability, and also for its properties of anti-inflammatory and clarifying.

Beta Carotene is removed for use from the carotenoids in the algae. Talking about the medicinal properties, algae carry Omega 3s, antioxidants, and it is proved to be anticancer and antiviral. Many of these properties of algae are evident as it is mentioned on ingredient labels of hundreds of products which people buy. It can also be put to use for making bioplastic, such as biodegradable water bottles or utensils that can be recycled along with the biofuel.

The biggest advantage of all these algae is carbon sequestration. Ocean-based algae is responsible for 45% to 50% of carbon dioxide consumed by the biosphere. However, they are small, their short life cycles make them structured and coherent. Research has proposed that algae powered bioreactors are 400x times more potent as compared to the trees at absorbing carbon dioxide.

1.3.4. Sea Cage Farming

Cage aquaculture inculcates rearing and nurturing of fish that emerge in circumscribed structures which permits the free interchange of water. Nonetheless the farming of fishes which occurs in cages have been exercised in the South East Asian countries since late 1800s, this technology of fish farming acquired admiration, demand, and popularity amidst the fish farmers and fisherfolk in India just a few years ago.

Cage farming provides enormous range for enhancing and amplifying the fish production in the country from the aquatic, estuarine, and coastal water bodies of the country. The wide-ranging water resources and favorable environmental conditions that are present in the coastal waters are exceptional for comprehensive cage farming.

Multispecies cage farming can be established in a commercial manner along with lowest investment in the country. Cage farming can be regarded as an extremely money-making alternative profession by the fishermen either discretely or in groups (Figure 1.4).

The relatively reasonable investment cost at beginning and recurring expenditure, the dexterity of operations, short time span of culture and inflated returns per unit volume are the reasons why cage aquaculture is a broadly accepted farming practice. In successful adaptation of any sort of farming technology, the existence of economy plays a significant part.

Figure 1.4. An illustration of sea cage farming.

Source: Image by Flickr.

In cage farming as compared to shore-based farming practices the finances like the investment and expenses incurred at the commencement are analogous with the expansion, the repairment of infrastructure are low gross income which is discerned by the mode of cage farming is excessive because the fishes that possess high value have substantial demand in the markets be it domestic or the export markets, both are cultured.

There are a lot of market opportunities for cage farmed fishes both in the domestic and export sectors owing to its pompous quality and pureness. Economic usefulness as well as financial accessibility review assists the farmers and fisherfolk for carrying forward decisions on investment and efficient administration of resources that are inadequate and sparse in order to obtain the economic benefits and profits at larger-scale.

The economic signifiers also help in progressing lending policies, reimbursement options or subsidies to the sector by policy makers or financial foundations.

1.3.5. Long Line Farming

Longlines are made use all over the globe in order to grow everything from scallops to seaweed. These systems are favored in regions that have high exposures. They are high yielding and elastic enough to tackle with the diversification of shellfish species along with the dimension of culture systems. In the process of longline system, the farmer anchors a length of line at both ends, connects a flotation, and suspends culture systems on the line.

Nursery rearing or nurturing and in addition to this grow-out can be achieved on longlines. Trays, tube modules and bags or cages can be dangled deeply in the water for implementing nursery rearing of clams, oysters, or scallops. Seeded lines or socks having sufficient predator protection are usually hung from longlines. Scallops are often grown-out on sunken longlines, in hanged lantern nets, or ear-hung straightaway on a down-line (Figure 1.5).

Figure 1.5. An illustration of long-line farming.

Source: Image by Wikimedia commons.

Layout of a longline system relies upon the features of site. The most important feature, from a viewpoint of security and consistency, is the presence of shore to moor one end of the longline. In few of the locations, both ends can be fastened and secured to the shore. Anchoring both the ends together in deep water might be done at sites where shore anchoring is even out of the question or desirable.

1.3.6. Raceway Farming

Explaining raceway in its most simple and uncomplex, form is just a flume for transferring water. Raceways made for fish culture are basically tanks which are comparatively schematic and depend on an elevated water flow in fraction to their volume with an intention to support the marine life. Flow-through fish culture systems drive water along the systems once, caters to waste treatment as needed, and then eventually dispense the water instead of treating and redistributing it. For making progression in aquaculture, the incursion of water meaning to say the water that flows inwards must be enclosed in the forbearance of temperature of the species that are being cultivated or cultured and should go with the quintessential temperature for the target species as firmly as possible.

The water that inflows also provide oxygen and is detached by the fish as the water escalates downwards in the raceway. In nearly all raceway systems, diffused oxygen is filled up and recharged by the water to plunge into subsequent tanks inside the raceway.

The metabolites that are dissolved from animals in the system are flowed out in the effluent, on the other hand, the particulate or particles waste that can be settled, are often apprehended by settling or sporadically by other methods of filtration.

Based on the water chemistry, the consumption and exhaustion of oxygen and the accumulation of ammonia, carbon dioxide, or fine particulates can become restricting to the production of fish within the system, over a period of time.

There are no natural foods that are produced in these systems, and diets which are nutritionally complete are mandatory for making the raceway aquaculture grow and expand and also it is essential for making it successful in every aspect.

Flow-through aquaculture systems needs the interchange of water to perpetuate appropriate water quality for the production of fish and depend on the flow of water for assembling and removing metabolic wastes. Water for flow-through facilities is most commonly channeled from streams, springs, or artesian wells to transfuse through the farm by gravity.

Water that is driven from wells or various other sources is way too costly and is used rarely. Water that is drawn from springs or surface sources for flow-through aquaculture is controlled by several public agencies, replying on the particular water laws of each state.

Deviation of surface water is regarded as a non-consumptive use, however, pumping groundwater from a well is reviewed as a consumptive use in some states. The release of a high-volume, dilute effluent from flow-through aquaculture provisions restricts the treatment options to a great extent which is available to producers from both technological and economic viewpoint.

Flow-through systems are the generally utilized aquaculture production systems for the culture of rainbow trout Oncorhynchus mykiss and other salmonids in the United States. Talking about the other Coldwater fish species that is generated in flow-through systems incorporate brook trout Salvelinus fontinalis and brown trout Salmo trutta. Flow-through systems put to use for production of freshwater phases of salmon.

Flow-through systems are also utilized on a bounded scale for the manufacture of warmwater fish for instance catfish *Ictal urus* spp. and tilapia *Oreochromis* spp. In the recent times, flow-through systems have been implemented and employed to produce cool water species like the yellow perch Perce flavescent, hybrid striped bass *Morone* spp., and numerous species of sturgeon *Acipenser* spp.

Flow-through systems incorporate linear earthen and palpable raceways and tanks established from other mediums and materials. Concrete raceways are the ones that are mostly used. Circular rearing tanks are also put to use in flow-through systems, usually for the production of bloodstock.

The quintessential raceway production system comprises of a tank, i.e., rearing unit or a pattern of rectangular tanks with the flow of water along the long axis. In an absolute raceway, water flow will approximate plug flow with consistent water velocity the tank cross segment.

Although, friction losses at the tank-water and air-water layers present at the borderline will result into water velocities to differ and range across the width and to the bottom of the raceway. Significant water velocities are at mid-depth, with moderately lowered velocities at the interface of air-water and greatly reduced velocities are experienced along the raceway bottom.

A defining and distinct characteristic of linear-pass raceways is a water quality gradient starting from the inflow towards the outflow of the rearing unit throughout production, along with best water quality at the inflow and worst water quality along the extent of the raceway as water flows toward the channel. Circular rearing units are more rigorously assorted and have comparatively uniform environmental conditions all along the tank.

In contrast to ponds, raceways have numerous perks and benefits. Per unit of space, raceway production is way too higher. Raceways also provide a much greater ability to examine the fish. This can make feeding more effective, and problems related to different diseases are easier to detect and observe at earlier stages. Once the disease signs and symptoms are examined, disease treatments in raceways are easier to implement and need fewer chemicals as compared to indistinguishable number of fish in a pond (due to the higher density in the raceway).

Raceways also permits closer monitoring or a detector of growth and mortality and superior inventory evaluations than ponds. Management inputs for example size grading are much more attainable and achievable in raceways than they are in ponds, and the harvesting becomes an easier task to do.

The drawbacks of raceways are principally associated to their requirement for large constant flows of uniform, high-quality water. Since such resources are rare, locating, and conserving a proper water supply is a vital component which should be focused critically.

Also, commercial viability at times requires that the water gravity flows through a series of raceways prior to its released. This adds on an obligation for an elevation or in other words a step up of the water source and appropriate topography for the gravity flow which is likely to take place between raceways.

Stating further, another limitation compared to ponds is the discharge of effluent. On one hand ponds substantially operate wastes enclosed in the culture systems, on the other, raceways, with their short retention times, do not. Effluent that are released from raceways are a greater and crucial factor, in comparison to the effluents discharged from the ponds.

1.3.7. Fish Hatcheries

In the dimension of food production, the industry of aquaculture is one of the fastest-growing and flourishing sectors, with approximately 40% of the total fish that result in as a source of food for humans that are produced in hatcheries.

Hatcheries are nearly similar to farms and are also identical to laboratories. The work of hatchery workers is to get the marine species spawned, hatched, and also, they monitor them. Once they grow big in size, the fish or shellfish are shifted into a farm or most of the times they are emancipated into the wild. Hatcheries are experiencing rise in their popularity because they account for providing an authentic source of stock replenishment for the waters of the world which are day by day becoming extremely over-farmed or over cultivated (Figure 1.6).

Figure 1.6. Harvesting fish from a hatchery.

Source: Image by Wikipedia.

The last decade has experienced a sudden rise when it comes to making the use of hatcheries due to the numerous factors, for instance, the increasing demand for various food products as the population keeps on constantly expanding, the impacts of over-fishing leaving and reducing the populations of natural fish, and the influences of the continuous change in the climate affecting the aquatic ecosystems.

While hatcheries play a major role in amplifying the food sources, yet they also come with major drawbacks. Day by day they are becoming increasingly controversial and contradictory because of the destructive and harmful impacts on the environment, health of human beings, and fish health that they have been largely interlinked with.

1.3.7.1. The Benefits of Fish Hatcheries

There are several perks and benefits of fish hatcheries that have made them so popular and has increased its demands. To start with, the outcome of the rising demand for food, the oceans and seas have progressed towards being intensely farmed.

Hatcheries permits for the replenishment and restoration of fish in these waters, at a much rapid degree as compared to the replenishment done by natural resources. This gives opportunities to the consumers to acquire access to a guaranteed stock and supply of seafood, along with this it ensures farmers that they will carry generating revenue and income from farming the animals. Hatcheries also have the abilities of offering food sources to carnivorous fish, validating their constant supply.

In the present decade, there have been a continuous increase in popularity of the diet which consists of seafood because of its ample health advantages as stated by several scientific studies. In the recent years, protein, and Omega-3 has become the focal point of people's diets to a large extent as it provides and helps in muscle growth, proper functioning of memory, eye health, reduced risk of various heart diseases, and many more.

In these materials and substances seafood is naturally excessive, which are favorable and useful not only to the general population but also it is responsible for having an influence on communities where malnutrition is the major problem, assisting people obtain access to nutrient-dense food. Hatcheries can guarantee a continuous supply of this food source which is significant.

Many cultures are extremely dependent on the industry of aquaculture to offer jobs in regions where poverty and lack of jobs is widespread. In China, for example, fish hatcheries implement approximately 4.3 million workers, making it a crucial industry which benefits and serves a larger part in the country's economy.

The process of running fish hatcheries is very labor-intensive and demands a lot of manpower, therefore, the industry can make a significant impact on employment rates in a community and provide an important boost to their economy.

Ultimately, in Europe some fish hatcheries are devoted to preserving scarce aquatic species which is serving and assisting to preserve them from vanishing from ecosystems all over the region.

1.3.7.2. The Drawbacks of Fish Hatcheries

While there are several advantages to fish hatcheries, it can be asserted that the stumbling block counterbalance these benefits. Firstly, fish hatcheries have been pinpointed as a wellspring of pollution to marine environments because of some of its practices which prevail.

In addition to this, due to the farming of these fishes in very intense conditions, their waste products turn out to be more strenuous, which can assist in the growth of microorganisms that can have a direct detrimental effect to the fishes.

Also, while the outlook of farmed fish is that they are cleaner as compared to the wild fish, Studies have suggested that the fishes that are famed have sky-scraping extents of toxic substances for instance antibiotics, organic pollutants, and pesticides in contrast to the wild fish.

The farmed fish that are being consumed, thus are at risk of the negative health impacts connected with consuming these impurities. Pregnant and nursing women are of specific risk to these adverse and harmful health effects.

Research have also proposed that fish belonging to the hatcheries possess less reproductive capabilities than wild fish. This can result into a big problem when hatchery fish are instigated into wild populations because it might be having a harmful influence on the fitness and wellness of the entire population.

Disease is also extensive in fish hatcheries that are not managed well and are organized poorly. For instance, Lice are varying usual in these environments, which can influence the quality, standards, and health of the fish. Diseases that are generated in fish farms can are responsible for effortlessly and simply spreading to wild fish and also accounts for putting the entire population of the wild wish in danger by contaminating them.

Eventually, it is identified that fish feel and go through pain and they can also experience stress. It is the nature of hatcheries to raise fish in very compact, narrow, and, stressful environments and slaughter them in a vicious way, either by asphyxiation, evisceration, or by starvation.

Collectively, there are numerous disadvantages that are considered important and is connected with hatcheries. While this technique of farming assists in maintaining levels of a food resources that are significant, provides jobs, and helps in coping up with malnutrition, the activities that are involved in hatcheries are in desperate need of streamlining in order to preserve the environment, human health, and highly regard the lives of the fish that are being farmed as well as it should be prioritized.

1.3.8. Integrated Multitrophic Aquaculture (IMTA)

The objective of Integrated multi-trophic aquaculture (IMTA) is to focus on the integrated production of aquaculture species of various trophic levels which comes under the dimension of a circular economy, reducing energy losses and degradation of environment (FAO, 2009; Chopin et al., 2012; Hughes and Black, 2016; Buck et al., 2018).

Under the production that is carried out in IMTA, the leftover or untouched feed and wastes of one species are retaken and transformed into feed, fertilizers, and energy which cater the needs of another species. IMTA can nurture and advance aquaculture sustainability, along with environmental, economic, and social benefits.

This can be attained by the medium of nutrient cycling, elevated economic flexibility emerging from growing production effectiveness, the diversification in product, and potential price premiums (Chopin et al., 2012; Van Osch et al., 2019).

The IMTA systems comes with various multiple layouts, amalgamating the production of vertebrate and invertebrate species as well as macroalgae. The organisms that are cultured and cultivated are fed aquatic species, such as fish or shrimp, and species drawing out the organic and inorganic matter from the water.

Species withdrawing the organic matter may be mussels, oysters, clams, sea urchins or polychaetas. These species rely for their food on organic waste, that is the uneaten food and feces. Species extricating the inorganic matter, for example the macroalgae (i.e., species of the genera Ulva, Gracillariid, Saccharina, Laminaria), seize, and make use of the wastes that are basically the inorganic nutrient.

IMTA permits the formation of more viable production systems because wastes of fish/shrimp production are not taken for granted rather they are held valuable as a resource instead of being considered a burden or pollution. This contributes to environmental sustainability and a more effective use of resources, while advocating the economic heterogeneousness (product diversification, bringing consistency in company by reducing risk factors), and social acceptability (best management practices).

Moreover, the IMTA has been identified as a contributor which plays a vital role in lowering the public opposition toward intensive aquaculture (Riddler et al., 2007; Alexander et al., 2016b; Buck et al., 2018).

However, there is still a pressing effort and requirement to enhance the awareness and spread information in the society, the ideologies, notions, and acceptability of aquaculture products, and propagate sound and accurate information to consumers regarding the aquaculture industry and its environmental sustainability.

Activities that encourage and promote understanding about sustainable aquaculture practices, the advantages of its products to consumers, and how they meet end-users' requirements are also mandatory to apply and execute stable and rapid consumption of aquaculture products.

In this respect, the concept of IMTA, and its benefits over the traditional methods, offers an exceptional opportunity to teach and educate students, and enlighten the general public, regarding the sustainability of environment, elevate their ocean literacy, in addition to this, it also ameliorates and enhances the acceptance and use of aquaculture products socially and globally, by amplifying its consumption (Shuve et al., 2009).

1.4. STRUCTURES FOR MARICULTURE OPERATIONS

There are four main varieties of structures that are in used prominently in the recent times. They are:

- Floating cages;
- Net enclosures;
- Earth ponds; and
- Continuous water circulation systems.

In the open sea or extensive shielded bays, the use of floating cages is done mainly for cultural objectives. Additional structures are not exercised due to the persistent action of waves. But in the context of culture in cages because of the expensive cages it is mandatory to sustain a plenty of individuals on the inside of the cage in order to make the venture beneficial and mainly profit-making.

Sequentially, this entails supplementary nourishment considering the fact that the amount of natural food which is present in the volume of water inside the cage is insufficient for maintaining the large number which are usually cultured in these types of cages. Apparently for this kind of culture practice to witness higher demand it will thus become essential to first develop cages at low costs and also fish feed at fair prices and understandably as a consequence, this technique and method has been proved gradual to spread its wings in developing countries.

Net enclosures act as a barrier off immense areas in sheltered bays are being implemented on a commercial aspect in a few countries. But a substantial amount of capital disbursement is essential and replacement done often times, which results from corrosion. Moreover, the additional feeding becomes significant which necessitates running costs which are excessive-but it does indemnify its path, this is the reason why in the advanced and industrialized countries some of the private farms are still existing.

Talking about the inland brackish water areas which have sufficient tidal range, have experienced in developing countries in the east has portrayed that in terms of mariculture, earth ponds established to confiscate spring tide water is said to be the most appropriate structure and framework.

This can be carried out without having much trouble and challenge in areas where the land is approximately flat or possess a slender slope along with the nature of the soil is up to the par. The construction of earth ponds and utilization of tidal water enhances the use of manual labor for the construction purposes and tackle with the gravitational force in the form of tides for the supply of water.

Apart from being self-sufficient and independent of any source of foreign imported energy it is also held accountable for the reduction of the imported material constituent to insignificant quantities.

The ceaseless water circulation units are accessible in some of the developed countries. These are the layout of large cement structures. However, along with this it also needs constant pumping and transport of water in plentitude and also the additional feed. Therefore, for this system an immense capital outlay becomes essential which also incurs high amount of repetitive costs. This showcases the economical aspect where a high price can be acquired for the final products and adequate raw material becomes accessible and available for feed.

1.4.1. Requirements for Coastal Mariculture Development in Earth Ponds

1. **Water Supply:** Tide is the most used and conventional source of water supply when it comes to coastal ponds. For this the tidal proportion must be appropriate for rapid changes of water in the ponds that occurs often. For providing better results, the perfect conditions must be to have 75 cm to 1 m of water above the ground at high tide whereas the water level at low tide should go down at the minimum of 75 cm below the existing ground level for the sake of facilitating the absolute evacuation of the ponds. While these are the prudent levels however, successful operations are implemented in parts where there is lower tidal amplitude, the reduced level is enough for carrying this out.

It is crucial that the supply of water should be eco-friendly and ideally of pH which range from 6.5 to 8. The requisite depth of water which should be available in the pond must depend on the following factors.

i. The species that is to be cultivated and cultured;

ii. As to whether it is to be:

a. a widespread system that is relying completely on the primary production which occurs naturally enclosed by the pond for food; or

b. semi-intensive using fertilizers; or

c. intensive system where the organisms cultivated would be provided supplementary nourishment that is the additional food.

2. **Soil Conditions:** For the earth ponds, the accessibility of soil containing good water holding capacity is significant so as to restrict loss of water through drainage or the leakage particularly when it is suggested to use fertilizers in order to raise the production. It is also important that the pH is within the desired boundaries as for the water given above. Furthermore, the use of acid sulphate soils should be avoided as much as possible.

3. **Seed for Stocking:** Availability of sufficient quantities of seed of the species to be cultivated is a key factor. If ample quantities are feasible naturally it would deduct the expenses of production to a reasonable extent. The production of seed which is done in hatcheries is being flourishingly implemented and executed for certain species in other countries. This might be regarded in situations when the natural supply is sparse and the demand for the seeds produced in hatchery is sufficient and satisfactory.

1.4.2. Culture Methods in Earth Ponds

There are mainly three methods, which additionally represents moderate progression in intricacy both in management as well as in financial outgoings and certainly associated with higher production. They are as follows:

1. **Extensive:** This is one of the easiest forms of culture where the use of fertilizer or additional feed is not required and the production carried out from the pond relies absolutely on the principal production which sequentially is based on sunlight and nutrients that are present in the water as well as in the soil.

The stocking density will thus have to be low and the repetitive costs are the cheap. For this technique, in the case of mullets, milkfish, and other herbivores or penaeids it is indispensable that the level of water must be maintained between 35 and 40 cm in the general platform area of the pond in order to uplift and support the growth of algae underneath the pond so that the organisms must feed upon and get their sustenance (Figure 1.7).

Figure 1.7. An illustration penaeid prawn.

Source: Image by Flickr.

If the depth of the water is profound then the intensity of light that outreaches the bottom is not adequate enough to promote the algal growth. In the tropics if the water level is decreased below 35 or 30 cm then the temperature is probable to rise and might also reach to limits that may be destructive.

2. **Semi-Intensive:** The next step with regard to the higher production is to stock at a slightly increased density and subsequently add fertilizers (which can be organic or inorganic) and in turn escalate the harvest from the pond. The role of the fertilizer is to expand primary production within the pond which is responsible for starting the chain reaction and results in higher production from a unit area of the pond.

   This method not only needs a bit of expensive financial functional cost, but it also inculcates supplementary labor for the gathering, stocking, and application of fertilizer. In developing countries, the farmers that are experienced at times fertilize their ponds due to generating income and the higher profit margin.

3. **Intensive:** In this mechanism the additional feed is added to the pond either in the presence of or absence of the fertilizers. The stocking rates would be increased to a great extent and along with this the depth of water in the pond is substantial as compared to the former cases because the food for the cultured organisms is not supported completely or is not based totally on primary production which takes place within the pond.

1.4.3. Environmental Considerations

Mariculture in earth ponds is restricted to the intertidal zone and usually to areas where the depth of the water at spring tides is not more than a meter.

In these areas appropriate for mariculture no fishing takes place and mariculture does not interrupt with the normal fishing practices. But in these areas in most parts of the Kenya coast mangrove vegetation is profuse. The mangrove is of numerous sizes.

In some areas such as Midas Creek the mangrove is entirely thick and is cut on a regular basis for various purposes like firewood, housing, and fencing. Where the vegetation is large-scale and of heavy timber, the costs of construction of earth ponds are way too high and this should be paid greater attention prior to picking up the areas for the growth and development of mariculture.

So far it is very soon to remark on the comparative revenue per unit area for mariculture and mangrove. The consequences of the pilot project which is under progression would offer it a rational comparison. In other areas for example north of Malindi, most likely as a result of semi-arid circumstances the supply of mangrove vegetation is short and as tracing forwards that is north from Malindi, it gets very bushy and scrubby.

Due to the inferior quality timber mangrove cutting is very much bounded near Ngobeni and as going further towards the north it is nil. But in areas such as the Tana River estuary which is the fresh water areas it is mainly tall grass wetlands which can be put back by mariculture farms.

Accepting the fact that the production of mariculture is more beneficial it is also significant to keep in consideration the influences of complete withdrawal of even the traces of mangrove from the region for the sake of establishing new ponds. While talking about the normal cycle of growth, the mangrove grows by consuming nutrients directly from the ground as well as water and produces roots, leaves, and stems.

In return to this the leaves fall into the water and perish there, improving the quality of the water. Frequently the stem is drawn out by man. The roots such as the leaves decompose and refines the ground and water. In the living state, the leaves of the mangrove transpire which causes the rise in humidity of the air.

When the mariculture farm puts back the mangrove vegetation, a high spring tide water is captured in earth ponds and it is reserved there in order to encourage and promote the development and enlargement of a carpet of algae on the bottom of the pond and phytoplankton which in return expands the production of crustacea, annelids, rotifers, etc.

Which caters to the nourishment of the organisms in the form of foods that is cultured in the pond. To put it in another way, construction of ponds for large-scale and semi-intensive systems do capture energy from the sun that is solar energy which eventually results into animal protein for the intake or consumption of humans. Moreover, the water which is present in the ponds keeps continuing to evaporate in place of the moisture that transpires from the mangrove leaves.

Salterns come under the other industries which are placed near the areas that are appropriate for mariculture. But since the fact that they rely on complete and sudden evaporation for which drier the soil conditions in an efficient way, they choose areas that are placed at higher altitudes in contrast to what can be made use for the different purposes of mariculture and there exists is no conflict or competition for the identical location.

In the initial stages of the growth and development of mariculture it is suggested to reply on the available seed in the natural waters in the absence of hatchery meaning to say it start off without the hatchery. The challenges that are demonstrated is that gathering of such large amount of post larvae and juveniles may impact the natural stocks and the normal commercial fishery in both the dimensions that is the lagoons as well as in the open sea.

Penaeids like most other marine fish are responsible for spawning large quantities of eggs-*P. japonicus* is known to spawn around 300,000 to 400,000 eggs each time and more than once every season. Yet eventually, a pair appears to get through and make it to maturity in order to spawn again.

Or else, they would experience an increasing population and it goes on expanding. By numbers the proportion which is made use of or exploited by the commercial fishery is slight as compared to what is lost to predators during the initial stages of the development when they are supposed to fall an easy prey.

In the early stages of development precisely during the time of its infancy, the mariculture activity having a little acreage would not be in a condition to be engaging or competing with the predators. Possibly some idea of the extent of predation can be acquired from the fact in a recent experiment in a pond of 1.1 ha area where there was not any availability of predators excluding for about 10 young snappers.

When this pond was evacuated except for those which could not be assembled with 5 mm × 5 mm mesh hand nets and what could go through 4.5 mm × 4.5 mm metal screens at the outlet channel an approximate of 72,000 four to six weeks old juveniles remained alive.

It is challenging and not so easy task to evaluate the number that are incapable of being collected or get free across the outlet gate screen and all these set their foot into the pond with the spring tide in the course of the exchange of water in the pond and were able to sustain and live because of the absence of the predators.

But such a large number is very exquisite in the same region under the prevailing natural conditions where predators are present. This gives a closer description of how much is lost or become a prey to predators under the existing open natural conditions.

Eventually, it may be suggested that what mariculture exercises does is to give a sanctuary for the desired and wanted species from their predators and to deal with escalating survival rate encourages primary production which rises to result into sources of food and nourishment for all these cultured organisms harnessing and regulating solar energy in the pattern or form of sunlight as well as the gravitational force in the form of tides. Therefore, mariculture would assist in raising production of food production and establish employment opportunities in the rural or underdeveloped regions.

1.5. THE MARINE ENVIRONMENT AND STRUCTURES FOR CULTURE

In the recent times, the term "Mariculture" can be also defined as the culture of organisms, which includes both plants as well as animals, in a marine medium or in other words in the aquatic environment which is either completely marine that is sea or the sea water mixed along with the fresh water to different extents.

This specific definition would inculcate both the areas of water that is the sea and inland brackish water areas. The problems and challenges faced in these regions for the purposes of aquaculture are fairly non-identical, entailing various approaches and dimensions to resolve them.

For instance, in the sea or aquatic environment, the motion of the water is continuous, and this happens due to tides, currents, and actions of the wind. This does assist in oxygenation and dispensation of the numerous elements, but it also does cause substantial and significant troubles and problems in building up the enclosures or restricted area for carrying out the culturing or cultivation of the organisms.

The familiar "ponds" which is generally associated with aquaculture establishing countries is not a pragmatic notion in the sea. Culture then has to be in a dissimilar form of enclosure and the type that are in popular use are cages anchored in bays that are protected and secured.

This may exist in a form of a metal frame containing netting material establishing the walls or the metal screens are used for building up the entire cage with or without a coating of corrosion resisting material. During the recent times the stiff synthetic material has replaced the metal but for this a metal frame work is essential. Although the cost of netting is low, it comes with its own drawbacks. It can be vandalized easily with the loss of organisms that are cultured.

In cages it is essential to sustain a great deal of individuals inside the cage in order to make the venture profitable and beneficial. This sequentially entails additional feeding, since the quantity of natural or the available food in the volume of water that lies inside the cage is not capable of maintaining the larger number generally cultured in these cages.

Where cages are used for fish culture carnivores are usually cultured in them. But the culture of herbivores is more popular and is commonly practices in the developing countries this is because of the insufficient supply of high protein fish feed which is necessary for species that are either predatory or carnivorous.

The popular cultivated herbivores do not do well or live fine in cages. In addition to this, the cost of cages is way too high and also need huge amount of capital investment which incorporates a high foreign exchange constituent. For these various reason's development and growth of culture practices in an aquatic environment has been sedate and is not progressing at an expected speed or in other words not taken off at all in developing countries.

Thus, to set foot in this field it is required to develop these two necessary items:

- Cheap enclosures which can resist and hold onto against the actions of the wave and occasional storms in the sea; and
- Cheap high protein fish feed which would be accountable for making cultivation of carnivorous species profitable.

In developed countries, additionally the culture in cages in the sea in secured bay area, they also practice generating seed in abundance in hatcheries and constantly seeding the natural fishing grounds thus promising recruitment.

Success has been evident in this field. The natural waters which are stored are renowned rich fishing grounds. Moreover, these countries were held accountable to take the danger and risk and foot the bill without guaranteed success in the course of the early stages of research and development.

In addition to being equally significant to the favorable outcome and success of this perspective and approach is the actuality that they were in a position to have treaty agreement with the other nations exploiting and consuming the same fishing grounds consequently being assured of an equitable and fair return for their investment.

These practices presuppose and involve not only massive expenditures and prices both in the form of workforce or labor and capital but also it incorporates the existence of natural rich fishing grounds and reasonably developing countries have been moving slow and steadily in the advancement and progression in this discipline.

1.6. CONCLUSION

This chapter gives an insight to the reader about the detailed overview of mariculture. It goes on discussing the trends in aquaculture, its potential for growth in wild fisheries as well as the production technology which emerged from Egypt and China. Further, it throws light on the various types of mariculture which includes the seawater ponds, tank farming, algaculture, sea cage farming, long line farming raceway farming, fish hatcheries, integrated multitrophic aquaculture.

As the chapter proceeds, it puts emphasis on the structure for mariculture operations. It starts discussing the requirements for coastal mariculture development in earth ponds, culture methods in earth ponds, environmental considerations.

Towards the end it illustrates the marine environment and structures for culture where it talks about the different definition of mariculture and how it is viewed in different frameworks.

REFERENCES

1. Aswathy, N., Imelda, J., Boby, I., & Shoji, J., (2020). *Economic Viability of Cage Fish Farming in India*. [online] Eprints.cmfri.org.in. Available at: http://eprints.cmfri.org.in/14274/1/CMFRI%20SP%20No.134_2020_Economic%20viability%20of%20cage%20fish%20farming%20in%20India.pdf (accessed 10 September 2021).
2. BCSGA, (n.d.). *Longline System | BCSGA*. [online] Bcsga.ca. Available at: https://bcsga.ca/shellfish-farming-101/shellfish-aquaculture/longline-system/ (accessed 10 September 2021).
3. Correia, M., Azevedo, I., Peres, H., Magalhães, R., Oliva-Teles, A., Almeida, C., & Guimarães, L., (2020). Integrated multi-trophic aquaculture: A laboratory and hands-on experimental activity to promote environmental sustainability awareness and value of aquaculture products. *Frontiers in Marine Science*, [online] 7. Available at: https://www.frontiersin.org/articles/10.3389/fmars.2020.00156/full (accessed 10 September 2021).
4. Imelda, J., (n.d.). *Background, Global trend, and Types of Mariculture* (pp. 15–23). [eBook] Kerala, India: Central Marine Fisheries Research Institute. Available at: http://eprints.cmfri.org.in/9705/1/Imelda.pdf (accessed 10 September 2021).
5. Kiara, C., (2020). *Algaculture: Our City Forest*. [online] Our City Forest. Available at: http://www.ourcityforest.org/blog/2020/4/10/algaculture (accessed 10 September 2021).
6. MESA, (2021). *Mariculture in Australia*. [online] Mesa.edu.au. Available at: http://www.mesa.edu.au/aquaculture/aquaculture02.asp (accessed 10 September 2021).
7. Sarah, M., (2020). *Fish Hatcheries-Benefits and Drawbacks*. [online] News-Medical.net. Available at: https://www.azolifesciences.com/article/Fish-Hatcheries-Benefits-and-Drawbacks.aspx (accessed 10 September 2021).
8. Sivalingam, S., (1981). *Basic Approach to Mariculture, Problems and Development Strategies*. [online] Fao.org. Available at: http://www.fao.org/3/AC571E/AC571E00.htm (accessed 10 September 2021).
9. Sivalingam, S., (1981). *Mariculture as a Method of Using Coastal Zones*. [online] Fao.org. Available at: http://www.fao.org/3/AC583E/AC583E00.htm (accessed 10 September 2021).

Chapter 2

Tropical Mariculture

CONTENTS

This chapter begins by explaining tropical mariculture, and it further highlights the farming of fishing in the tropics. It also goes on explaining the ocean fish farming in the tropics and sub-tropics that are most impacted by climate change.

It also sheds light on the integrated multi-trophic aquaculture (IMTA) which is often termed as IMTA in the marine temperate waters. It further elucidates the different concepts of IMTA along with explaining the IMTA system that prevail in the recent years.

Towards the end, it addresses the integrated tropical mariculture where it defines the systems classifications, the growth of tropical offshore mariculture and the status of offshore aquaculture in tropical regions.

2.1. INTRODUCTION

Mariculture is known as the cultivation or farming of marine organisms in their natural habitats, mainly for the commercial objectives. Mariculture is often referred to as the culture of organisms, including altogether plants as well as animals in both the sea and the areas of inland brackish water.

All over the globe, mariculture is expanding and as stated by FAO statistics, it depicted an elevation from 9 million tons in 1990 to more than 24.7 million tons during the year 2012.

Few of the species have shown prospective for sustainable growth in the coming years but might be tactful aquaculture efforts are *Acipenser* spp., *Anguilla* spp., *Epinephelus* spp., *Lates* spp., *Lutjanus* spp., *Oreochromis* spp., *Thunnus* spp., and *Ulva* spp. (CBD, 2004). Mariculture is influenced and controlled by seaweed (Japanese Kelp) and mollusks (Pacific cupped oyster) and finfish salmon that are of immeasurable value (Figure 2.1).

In addition to this, the small-scale cultures of Sea horse are present, that are giant clam, microalgae, rotifers, and brine shrimp. The species such as milkfish, etroplus, and mullets are cultured and cultivated in brackish water.

Concurrently the influence and pressure on the aquatic resources and wild fish stock is manifesting a growing tendency as the human population expands, basically they are directly proportional to each other. The Global marine catch was about 14 million tons in the year 1950, which increased to 65 MMT towards the beginning of 2012 (FAO, 2014).

Figure 2.1. An illustration of kelp.

Source: Image by Flickr.

It is also eminent that the aggregate catch was more or less consistent around 70–65 MMT over the last 25 years and implies that there might not be supplement increase in the coming years from the capture segment. Mariculture provides a better-quality food also it is more efficient and effective as compared to numerous other food production systems.

It is conceded that all forms of mariculture muddle the biodiversity at all levels, be it species, genetic, and ecosystem, which can be responsible for causing negative impacts. The primary effects results into the degradation of habitats, diminish in wild populations, establishment of species that are non-indigenous, biological pollution, genetic influences of target species and social impacts like the issues and factors related to human health, reduction in the job opportunities and shortage of employment income of conventional fishermen.

There are various distinct open dimensions and aspects for overcoming the negative effects of mariculture related to biodiversity. They incorporate the effective site selection, proper evaluation on environment, proper feeding protocol and concords, greater outflow and steps or measure to prevent waste discharges, better management and organization related to genetic resources, constructing hatcheries for the collection of seeds, reducing the collection of wild seed, and amplifying positive, pragmatic effect of mariculture in order to lessen the pressure on capture fisheries.

2.2. FISH FARMING IN THE TROPICS

Fish farming is the industry of producing fish by the means of husbandry or culture for food, recreation, and other distinct meanings or purposes, which marks its history to ancient Egypt and China. There are approximately 100 different species of fish across the globe that are cultivated and farmed.

Elaborating fish farming, it consists of breeding, rearing, and nurturing of young and emerge to market size of species which are cultured or cultivated. The key principles of the farming of fish's envelopes the ways fish adapt and adjust to the marine environment, their food habits, nourishment, sustenance, and breeding characteristics.

The essential components for growth and survival of the fish that is the availability of adequate oxygen in the water, food, and the removal of waste. The aquatic environment must remain clean and free of the more human interreference so as to keep the fishes healthy and also to keep them alive for the long duration of time.

On the other hand, in the tropics while farming the tilapias, as an instance, the culture units which are made use of and implemented are mainly ponds, tanks, and net cages. The methods of manufacturing differ in accordance with the management that is applied for example those for large-scale, semi-intensive and small-scale systems.

Fishes that are befitting for culture are the ones which are bred effortlessly and smoothly, show faster growth, are hardy and have an excellent market. Techniques and various methods for induced breeding of fishes, mono-sex culture, regulation of parasites and diseases, polyculture, and implying farming systems that are unsegregated are applied in the process of fish farming to amplify the supply, production, and productivity of seeds.

While playing the role of an enterprise, fish farming is expansive and is exposed to the same aspects and principles of economy as that of any industry which is completely dependent upon capital, labor, manpower, and various other resources.

As compared to other industries that deals and contributes in the animal protein production and fabrication, fish farming is held and considered as the most efficient, effective, and money and profit making and is capable of producing better results.

Along with the consumption and depletion of the wild fish stocks prevailing in worldwide, aquaculture is anticipated to likely produce fish in order to fulfill the demand that is growing rapidly. While talking about the future of cultivation and farming of the fish, it is said and expected to evolve brightly and in a better way in the coming days.

The growth and expansion of the farming of the fish in estuaries as well as in the aquatic areas rather than in freshwater bodies is predicted beforehand due to the growing demand for freshwater. The government policies and schemes for zoning fish farms, management of the practices and different methods for the preservation of the environmental and also the development of market is significant and required for sustainable fish farming.

While explaining the farming of the Fish, it is the business or industry that deals in producing fish by means of husbandry and has the same purpose as that of a fish culture. In the extensive terms, fish farming is the segment of aquaculture, which revolves around the culture of plants and animals that are aquatic.

The farming of fish consists of breeding, rearing of the young and the emergence of young fish to adult or fish which are harvestable. Fishes are cultured and cultivated in freshwater, brackish water, and seawater for acquiring food, enjoyment, refreshment, and different other purposes (Figure 2.2).

Figure 2.2. An illustration of tilapia fish.

Source: Image by Wikimedia Commons.

The history of Fish husbandry begins from 2500 B.C. when the Nile tilapia was kept in holding ponds in prehistoric Egypt. The reproduction of common carp that is found in China was recorded rapidly in 475 B.C. Carp culture was carried out on a broader scale in Europe in the Middle Ages somewhere around 1400–1600.

The pre-eminent achievement in the artificial fertilization of brown trout eggs was disclosed and documented in 1853 in the United States. Commercial culture of warmwater fish was started during the end of the 1920s.

For demonstrating the propositions and practices of the farming of fish in the tropics, tilapia farming is considered to be one of the best examples.

Illustrating the Tilapias, they are warm water fishes that emerge from the land of Africa. The paramount species of tilapia that are commercially farmed in approximately 40 countries of the world are the Nile tilapia (*Oreochromis niloticus*), *Mozambique tilapia* (*O. mossambicus*), and blue tilapia (*O. aureus*), and their hybrids.

A variant form of tilapia popularly known as the red tilapia is also cultured on a large scale. Tilapias are prudent and beneficial fishes when it comes to farming as they are bred in captivity without causing much difficulty, grow very fast and are profit making and economical to culture.

Likewise, food fishes, tilapias are also well undertaken and in various regions across the world mainly in Asia, Africa, and the Americas. The chief and major countries that accounts for tilapia are China, Indonesia, Thailand, and the Philippines. Tilapias sustain low in the food chain.

They are responsible for rapidly taking up the natural food in the water for example plants and animals (plankton)that are microscopic, by means of their systematic and structured nourishing apparatus but they are also responsible for accepting artificial diets. The fishes are specifically adapted and to digesting blue-green algal cells with the highly acidic conditions occurring in their stomachs.

The process of breeding of tilapias goes in such a way that the male constructs a nest at the substratum to attract the attention of the female. The male has the potential of breeding with more than one female, they are not limited to single female. The female tilapia mouth broods the fertilized eggs, which hatch into sac, and fry in the core of the buccal cavity.

The female delivers the fry after 7 to 10 days from hatching. Tilapias expands to maturity and it takes them less than four months to do this in the tropics. Breeding of tilapias can also take place throughout the year. Male tilapias grow faster as compared to the females and are specialized to grow up to a size of 3 kg or sometimes even more than that.

Tilapias are cultured mainly in ponds, tanks, and net cages in freshwater, brackish water or at times in seawater. The Nile tilapia (as stated in Figure 2.2) is the most significant species for freshwater culture on the other hand, the Mozambique tilapia is basically the most salt-tolerant.

2.3. OCEAN FISH FARMING IN TROPICS AND SUB-TROPICS MOST IMPACTED BY CLIMATE CHANGE

Diners might preferably discover many farmed oysters and lesser Atlantic salmon on their particular plates as transformations in climate keeps warm the pacific coast of Canada.

In a research study which was published in Global Change Biology, researchers at the University of British Columbia did a thorough study and discovered how climate change can affect 85 species of fish and mollusks that are generally and usually farmed or cultured in seawater.

In their study they also found out that certain species such as Atlantic salmon, European seabass and cobia, and particular regions like the tropics and the Arctic, could be specifically at risk due to the various effects of climate change.

Taking the example of the pacific coast of Canada: during the mid-21st century, the region is expected and forecasted to lose between 60% and 84% of its area which is at present is suitable for Atlantic salmon farming under the influence of strong lightening and dulling or mainly the mitigation, the greenhouse gas emissions which is low, and sometimes no mitigation, high emissions scenarios, which happens sequentially.

In comparison, the region is responsible for obtaining 46% more area for Pacific cupped oyster farming under the framework of high emissions cases in the coming future that is by the 2050s, whereas Norway and Sweden could respectively see gains between 48% and 100% in areas which are suitable and appropriate for Atlantic salmon farming.

As stated by the lead author Muhammed Oyinlola, "In total, higher latitudes are estimated to acquire nearly 40% more species, on the other hand, tropical to subtropical regions are capable of losing between 10% and 40% of species that are appropriate for marine aquaculture towards the 2050s," Muhammed Oyinlola, is a postdoctoral fellow at UBC's Institute for the Oceans and Fisheries, who was a PhD student with the Nippon Foundation-UBC Nereus Program when he conducted the study.

He also contended, "Tropical and subtropical regions rely the most heavily on seafood for food security, income, and employment, but our projections show that they are likely to suffer the greatest losses."

This research also assembled and recorded environmental data for example temperature, salinity, oxygen levels and all the familiar locations of the available and existing farm operations in order to evaluate the total number of species could in the recent times be farmed, cultured, and cultivated in various countries' absolute and unshared economic zone, that most of the times encloses waters in the dimensions of 200 nautical miles from the coast.

After this the researchers also highlighted how many species will be obtainable and accessible in the near future under the conditions of different climate or the climate changes.

Few of the world's smallest island nations and prevalent states in the pacific and western Indian Ocean, for instance, could record or face the decline in the number of species available for marine aquaculture by 100% in the time span of few decades under the conditions of high emission.

Simultaneously, Panama, which is a homeland to one of the world's largest open-ocean fish farm, could also experience a decline and near total loss in area in the respect of cobia farming—a species which is regarded as one of the best suited and relevant for offshore aquaculture.

"People are counting on aquaculture to produce seafood to meet the growing demand while capture fisheries have reached their limits or being over-exploited. However, this study highlights that fish farming will not be exempt from the impacts of climate change," stated by William Cheung, a professor and principal investigator of the study and Canada Research Chair in Ocean Sustainability under Global Change at UBC's Institute for the Oceans and Fisheries.

"While some northern regions such as Norway or Sweden are projected to gain more species and area for marine aquaculture, such opportunities pose new risks as well. Arctic ecosystems for example are very sensitive to disturbance from human activities," as stated and explained by Oyinlola.

Further as elaborated by Cheung, "Ultimately, the most effective ways to ensure our seafood security is to mitigate climate change and eliminate overfishing,"

2.4. INTEGRATED MULTI-TROPHIC AQUACULTURE (IMTA) IN MARINE TEMPERATE WATERS

The temperate zone of the globe usually denoted to the area that lies between latitudes 23.5° and 66.5° in both hemispheres (Milne, 1995). This incorporates oceanic waters in the temperature range of 7–25°C, even though in winter season the water temperature can fall down to the near freezing point in higher latitudes (Levington, 1995).

This report contemplates countries in this range, specifically Canada, Chile, Finland, France, Ireland, Norway, Portugal, Spain, South Africa, Sweden, the United Kingdom (mainly Scotland) and the United States of America-they all of have most functional aquaculture industries and either not on so larger scale IMTA systems already held in practice or at times hold potential for the growth and development of IMTA.

Illustrating IMTA, it is an implementation and practice in which the by-products that is wastes, from one species are recovered and reused and convert itself to works as inputs (fertilizers, food, and energy) for another.

Fed aquaculture species (e.g., finfish/shrimps) are gathered together, in suitable proportions, with organic extractive aquaculture species such as suspension feeders/deposit feeders/herbivorous fish whereas the inorganic extractive aquaculture species such as, e.g., seaweeds, for the management of proper and balanced ecosystem approach that takes into account the various aspects like site specificity, operational limits, and food safety guidelines and regulations.

The objective is to achieve and provide a sustainable environment through the process of bio mitigation, economic stability by means of the diversification of products and reduction in risks, along with social acceptability through the well-structured management and their practices so as to implement them.

Multi-trophic is known as the incorporation of species from the various trophic or nutritional levels belonging to the same system as contended by Chopin and Robinson (2004); and Chopin (2006). This is said to be the one potential difference that comes from the age-old practice prevailing in aquatic polyculture, which could plainly be the co-culture of distinct fish species from the same trophic level.

In this respect, these organisms, can sometimes share the same biological and chemical processes with each other, along with implying some synergistic benefits, which eventually leads to important shifts in the ecosystem. Several traditional polyculture systems might include a broad diversity of species, which inhabits several niches, as a large-scale culture that is low intensity, low management enclosed in the same pond.

The integrated in IMTA refers to the more intensive cultivation of the different species in juxtaposition of each other (but not necessarily right at the same location), associated by nutrient and transfer of energy which is done by water. Ideally, in an IMTA system, the biological and chemical processes must balance.

This is obtained by medium of the suitable selection and proportions of various species catering to distinct roles and functions of ecosystem. The co-cultured species should be more than merely biofilters; it is essential to have them as harvestable crops providing commercial value (Figure 2.3) (Chopin, 2006).

Figure 2.3. An illustration of IMTA in freshwater pond.

Source: Image by Wikimedia commons.

A working IMTA system should result in extensive production for the whole system, constructed on correlative advantages to the co-cultured species and ameliorate the health of ecosystem, despite the production carried out on an individual's basis of some of the species is lower in contrast to what could be achieved in monoculture practices over a short stretch of time period (Neori et al., 2004).

Many a times the most used term that is "integrated aquaculture" is made use to explain the amalgamation of monocultures by the medium of transfer of water in between the organisms as explained by Neori et al. in 2004. For all intents and purposes, although, both the terms that is "IMTA" and "integrated aquaculture" are primarily different from one another in the context of their degree of descriptiveness. Sometimes, these terms are swapped.

Aquaponics, fractionated aquaculture, IAAS (integrated agriculture-aquaculture systems), IPUAS (integrated peri-urban aquaculture systems), and IFAS (integrated fisheries-aquaculture systems) are also said to be the variations of the concept of IMTA. The IMTA concept is way too flexible.

IMTA systems can be land-based or open-water systems, marine or freshwater systems, and might consist of numerous combinations of species (Neori et al., 2004). Few of the IMTA systems have incorporated such as shellfish/ shrimp, fish/seaweed/shellfish, fish/shrimp, and seaweed/shrimp (Troell et al., 2003).

Here, the significant factor is that the suitable organisms are selected based upon the role they play in the ecosystem, their economic worth or potential, and their demand or acceptance by consumers. While IMTA likely happens because of the traditional or incidental, adjacent culture of varying species in few of the coastal areas (Troell et al., 2003), the IMTA sites which are purposefully designed are, in the recent years, very less common. In addition to this, in the present time are simplified systems, like fish / seaweed / shellfish.

In the coming future, more progressive systems with various other elements for various other functions, or in some cases similar functions but different size ranges of organic particles, will have to be drawn and designed (Chopin, 2006).

The objective is to expand long-term sustainability and profitability per cultivation unit (not per species in isolation as is done in monoculture), as the wastes of one crop (fed animals) are transformed into fertilizer, food, and energy for the other crops (extractive plants and animals), which can eventually be in return sold on the market.

Among this, Feed is one of the crucial and core operational costs of finfish aquaculture operations. By the mode of IMTA, several of the food, nutrients, and energy considered lost in finfish monoculture are recaptured and then transformed into crops which provides commercial value, while bio mitigation takes place. Hence, this way all the constituents of cultivation possess an economic value, as well as a prime role in services and recycling, reusing processes of the system, along with the harvesting of the three types of crops taking part in the export of nutrients exterior to the coastal ecosystem.

IMTA is regarded as the more sustainable than the common monoculture systems-meaning to say, a system of aquaculture where only one species is cultured-in that fed monoculture have the propensity to have an influence on their local environments which is caused because of their reliability of supplementation with an exogenous source of food and energy in the absence of mitigation (Chopin et al., 2001).

For 20 years from now, several authors have shown that this exogenous source of energy for instance fish food can have a substantial influence on organic matter and nutrient loading in the aquatic coastal areas (Gowen and Bradbury, 1987; Folke and Kautsky, 1989; Chopin et al., 1999; Cromey, Nickell, and Black, 2002), having an impact on the sediments that lie beneath the culture sites and production of variations in the composition of nutrients present in the water column (Chopin et al., 2001).

Association of various species in one culture unit can lessen these influences because the culture of the species that do not need exogenous feeding may set the system outputs at equilibrium by the process of energy conversion, by means of which the waste of one species becomes the food for another species as stated by Chopin et al. (2001).

To describe this with an example, the wastes given off from the culture of salmon, for example uneaten fish food, fish faces and eliminated nitrogen (N) and phosphorus (P), can be absorbed by shellfish (organic processors) as well as seaweed, i.e., inorganic processors, so as to lessening the amount of waste released from a fish farm and converting it into fodder for various other species which possess commercial value.

This implementation and execution of IMTA can assist in reducing adverse effects of the environment whereby also establishing other economically manageable products in the same framework, parameter, and time.

It provides binary advantages which should make IMTA appealing when perceived by the fish farmers, while establishing the system of aquaculture, more acceptable to environmentalists and also to the inhabitants in general.

Considering the potential for increased profitability, it is wonderful to comprehend how very little the aquaculture sector has different and is manifolded in some countries or in important producing regions. For instance, In Canada, the salmon aquaculture represents 68.2% of the haul of the aquaculture industry while 87.2% of its farmgate value (Chopin and Bastarache, 2004).

In countries like Norway, Scotland, and Chile, the salmon aquaculture shows 88.8%, 93.3% and 81.9% of the freight of the aquaculture industry, as well as while talking about the farmgate value, it is 87.3%, 90.9% and 95.5%, respectively (Chopin et al., 2008).

Contrarily, while Spain (Galicia), is responsible for manufacturing only 8% of salmon in tonnage (16% in farmgate value), it manufactures 81% of its freight in mussels (28% in farmgate value).

Having excessive of production intensive in a single species sets the business at risk, unsafe, and unprotected as well as exposed to issues of sustainability due to the low prices which results due to oversupply, and the possibility of natural disasters which causes damage at a greater scale especially to the one's only crop, i.e., diseases, adverse weather conditions.

Therefore, diversification of the aquaculture industry mainly which takes place at the local and regional levels is crucial to deduct the economic risks and perpetuate its consistency and competitiveness. The conventional perspective of diversification most of the time refers to producing another product in addition to the same lines of the first, that would be relevant into the existing production and framework of marketing.

In countries like North America and Northern Europe, in finfish aquaculture, this has generally meant salmon, cod, haddock or halibut. Although, from an ecological perspective, these are all "shades of the same color."

No symbiosis is established; instead, these situations amalgamate the influences on the system. Actual ecological diversification of aquaculture implies farming at more than one trophic level, i.e., shifting from one species of finfish to another group of organisms of lower trophic level for example shellfish, seaweeds, echinoderms, polychaetas, bacteria, etc., looking alike a natural ecosystem.

Holding on to the same ecological trophic level will not represent few of the environmental issues as the system remains unbalanced and deranged due to the non-stable dispensation of energy and non-diversified resource needs and results.

Diversification of products also stands for perceiving seafood from a different perspective. Aquaculture products on the market today are very alike to those which are acquired from the conventional fishery resources, and are, therefore, in frequent direct competition. This may be a section of the market forces at work, the prevailing opportunity is utilized to diversify and change from the fish filets, or mussels and oysters on a plate in a restaurant, to a wider untapped formation of bioactive constituents of marine origin for example pharmaceuticals, nutraceuticals, functional foods, cosmeceuticals, botanicals, pigments, agrochemicals, and bio stimulants, and industry-relevant molecules.

Accordingly, research, and development on different species should no longer be regarded as R&D on the substitutes of finfish species, but instead on alternative aquatic products. In addition to this, diversification should be observed and contemplated as an investment portfolio, with either short-term, or long-term, hazardous, or harmless components, while keeping an eye on the objectives that must include the long-term growth and consistency.

There arises a contradictory situation when glancing at current food production in the recent years all over the globe. In agriculture, 80% of the production consists of plants and 20% of animal products comprises of meat, milk, eggs, etc., whereas in aquaculture, 80% of the production is animal biomass and 20% deals in plant biomass (Chopin and Reinertsen, 2003).

While talking about precisely mariculture, the overall production in the world somewhere around 2004 consisted of 45.9% seaweeds, 43.0% mollusks, 8.9% finfish, 1.8% crustaceans, and 0.4% of distinct other animals (FAO, 2006a). Subsequently, in various parts of the world, aquaculture is not as similar to the finfish aquaculture, as believed by many people living in western countries who are well off.

Constructed on the requirement for maintaining the balance of the cultured species functions enclosed in the surrounding ecosystem functions, marine herbivores, carnivores, and omnivores ought not be cultivated or cultured while keeping aside marine plants-as efficient and effective biofilters, either a crop on their own, or a food constituent for other organisms-if people have to make the "Blue Revolution" (Sensu Costa-Pierce, 2002) that is obviously "greener." Numerous species of seaweeds that are cultivated and cultured in the light of the right conditions, mainly near sources of high concentrations of nitrogen as in accessibility to finfish farms, can be an exceptional source of proteins, significant amino acids, and unsaturated oils.

People need to have knowledge and be aware of the systems that comprises of production of other food items while dealing with the rest of the world if an individual wants to comprehend, their present prevailing system and accurate position it in viewpoint with other systems.

Seaweeds and micro-algivores for example filter feeding shellfish and herbivorous fish deals with 59% of the aquaculture production, all across the world, followed by the production of 30% of omnivores and detritivores.

In tonnage, the three distinct paramount aquacultured species are-firstly the seaweed Laminaria japonica, and secondly the two micro-algivores, the Pacific cupped oyster, *Crassostrea gigas*, and also the silver carp, *Hypophthalmichthys molitrix*.

Vocal public resistance to aquaculture has been established by "high value" salmonids and other marine fish and shrimp which are carnivores, even though, in fact, it represents merely 10.7% of the world's aggregate mariculture production (but 40.8% of its value).

From the above-mentioned numbers and records for mariculture, an individual may tend to think that at the world level, the two different types of aquacultures are balanced comparatively, one is the fed and the other is extractive.

Although, due to the primary approach of monoculture, these different types of production which deals in aquaculture are most of the time geographically separate, and, consequently, they balance each other only one in a while out on the local or regional scale.

For instance, in Eastern Canada, fed salmon aquaculture is essentially located in the Bay of Fundy in Southern New Brunswick and in Southern Newfoundland, on the other hand extractive mussel and oyster aquaculture is located in the Northumberland Strait and the Lower Gulf of St. Lawrence, through the coastlines of Prince Edward Island and Northeastern New Brunswick, and in Eastern Nova Scotia and Northeastern Newfoundland.

In Japan, aquaculture is in most of the cases carried out and implemented with abundant bays dedicated to either shellfish, seaweed, or finfish aquaculture. Nevertheless, there are instances in China of bays that are organized according to the IMTA approach (Chopin and Sawhney, 2009).

While IMTA may be viewed as a new concept to western farmers, this outlook of farming, cultivation, and aquaculture has long been in use mainly in Asian countries. Japan and China have used this strategy and technique for the co-culture of rice and fish for millennia (Neori et al., 2004).

Although the cultured species differ, why, then, is this common-sense solution not implemented on a large-scale, especially when it comes to the western world? The main reasons for this generally revolve around social customs and practices that one is ideally aware of even if common sense tells them that they should amplify them.

Human society does not transform rapidly unless there are reasonable reasons to. The restricted nature of the marine food production industries is a better example of the relative slowness with which such transformations are adapted and adjusted, mainly when dealing with a complicated aquatic environment, which is mostly seen only on the surface of, and have been hard to understand the processes and actions taking place beneath it over substantial distances and volumes.

Western countries constantly keep on reinventing the wheel. Various studies on integrated methods for treating wastes from modern mariculture systems began in the 1970s (Ryther, DeBoer, and Lapointe, 1978).

After that period, the scientific interest in IMTA got still, and it was not until the late 1980s and during the beginning of 1990s (Intergard and Jensen, 1983; Buschmann, López and Medina, 1996; Kautsky, Troell, and Folke, 1996; Chopin et al., 1999) that a renewed interest took place, which is fundamentally based on the common-sense approach that the solution to nitrification is not dilution but metamorphosis within the perspective of an ecosystem-based management.

This interest has likely been an indirect result of the increased demand for aquaculture products. In 2004, aquaculture production from mariculture was 30.2 million tons, representing 50.9% of the global aquaculture (FAO, 2006a), which has steadily increased each year since the 1950s, at a rate of roughly 10% (FAO, 2006a).

This increase has in turn, resulted in intensified cultures, decrease in available habitat (space available for cage sites/aquaculture leases), and increased environmental impacts on the immediate ecosystem. IMTA is a method whereby production can be intensified, diversified, and yet remain environmentally responsible-thereby ensuring a sustainable aquaculture industry.

Multi-trophic integration appears to be a logical next step in the evolution of aquaculture. This trend in the global recognition of the need for more advanced ecosystem-based aquaculture systems has started showing up in the scientific world by means of the aquaculture conference circuit.

For instance, in recognition of this growing interest, the Aquaculture Europe 2003 Conference in Trondheim, Norway, whose theme was "Beyond Monoculture," was the first large international meeting (389 participants from 41 countries) with IMTA as the main topic.

In 2006, at the joint European Aquaculture Society and World Aquaculture Society Conference in Florence, Italy, IMTA was represented as a serious research priority and option for consideration on the future growth and development of aquaculture practices and culture.

2.4.1. Concept of Integrated Multi-Trophic Aquaculture (IMTA)

Integrated aquaculture systems as described in detail by Neori et al., (2004); Barrington et al. (2009); and Angel and Freeman (2009) and case studies in India are explained below in brief. In most of the monoculture farming systems the fed-aquaculture species and the organic/ inorganic extractive aquaculture species (bivalves, herbivorous fishes, and aquatic plants) are farmed accordingly and independently instead of together in different geographical locations, which causes in pronounced shift in the different processes of environment.

IMTA incorporates cultivating fed species with extractive species that make use of the inorganic and organic wastes from aquaculture which assists in their growth. According to Barrington (2009), "IMTA is the practice which combines, in the suitable proportions, the cultivation and farming of fed aquaculture species (e.g., finfish/shrimp) with organic extractive aquaculture species (e.g., shellfish/herbivorous fish) and inorganic extractive aquaculture species (e.g., seaweed) to establish balanced systems for environmental sustainability (bio mitigation) economic stability (product diversification and risk reduction) and social acceptability (better management practices)."

This farming method and technique is way different from finfish "polyculture," where the fishes share the same biological and chemical compositions which could potentially give rise to the large shifts in ecosystem. Multi-trophic is the amalgamation of species from various trophic levels in the same system. The multi-trophic sub-systems are non-segregated in IMTA that refers to the more intensive cultivation of the different species in closeness of each other, associated by transfer of energy and nutrient through water.

2.4.2. Current IMTA Systems

The IMTA concept is extremely flexible. It can be applied to open-water and land-based systems, and marine and freshwater systems (sometimes then called "aquaponics" or "partitioned aquaculture"). What is important is that the appropriate organisms are chosen based on the functions they have in the ecosystem and, moreover, for their economic value or potential.

What is quite remarkable, in fact, is that IMTA is doing nothing other than recreating a simplified, cultivated ecosystem in balance with its surroundings instead of introducing a biomass of a certain type expecting this can be cultivated in isolation of everything else.

Moreover, IMTA goes beyond environmental sustainability; it provides economic diversification and reduces economic risk when the appropriate species are chosen, and it increases the acceptability of the overall aquaculture sector by using practices evaluated as responsible by the industry, the regulators, and the general public.

Presently, the most advanced IMTA systems in open marine waters have three components (fish, suspension feeders such as shellfish, and seaweeds in cages and rafts), but they are admittedly simplified systems.

More advanced systems will have several other components, best suitable examples are crustaceans in mid-water reefs; deposit feeders such as sea cucumbers, sea urchins and polychaetes in bottom cages or suspended trays; and bottom-dwelling fish in bottom cages, they can be used for either distinct or same functions or usage, yet for non-identical size brackets of particles, or can say, they are selected for their appearance at different times or seasons of the year.

2.5. INTEGRATED TROPICAL MARICULTURE

As compared to the many integrated systems that are present in freshwater aquaculture (being an integral part of agriculture) monoculture that is the intensive farming is the primary ideology in mariculture.

However, in the tropics a person may discover various deviations resulting from such generalization, mainly in brackish water pond systems.

Though it is not so difficult to envisage the interconnectedness of different species to cure and deal with effluents that makes the use of mollusks, seaweeds, etc., as biofilters by the method of using effective resource utilization, with only a few restricted number of intensive integrated farming methods or systems have been applied and executed in the tropics as stated by Phillips (1998).

Numerous literature reviews have gathered and structured information about these (Chien, 1993; Chien and Liao, 1995; de la Cruz, 1995; Lin, 1995, 2006; Gavine, Phillips, and Kenway, 1996; Brzeski and Newkirk, 1997; Lin and Yi, 1999; Browdy et al., 2001; Fast and Metasveta, 1998, 2000; Shyu and Liao, 2004; Neori et al., 2004).

Although, none of these tend to provide a thorough outline and overview, some of them are too extensive, while some focus on particular systems (shrimp farms) or are limited to specific countries. With hardly a few exceptions these reviews and studies emphasize on logistical obstruction to combination.

2.5.1. Systems Classification

The various forms of prevailing integrated mariculture systems are differentiated from each other on the basis of choice of species and the framework of design. A few ventures have been made for carrying out the categorization of such systems.

Hambrey and Tanyaros (2003) contended and focused on the large number of systems, and classified them in accordance with their own experiences from the field that is especially in Southeast Asia, under:

- Integrated ponds that is the polyculture fish, agriculture inputs-fish, while in;
- Integrated ponds and field systems that is intensive fish-extensive fish, rice-fish/shrimps, shrimp-oyster/seaweed, shrimp-mangrove, shrimp-fish and lastly in;
- Cage-open water systems that is fish/mollusks/seaweeds.

In 1995, De la Cruz pointed out three classes of brackish water integrated farming systems (BIFS) for Southeast Asia: they are: (i) aquaculture-agriculture; (ii) aquaculture-silviculture; and (iii) brackish water and marine polyculture.

He also identified that only the countries like Indonesia, the Philippines, Thailand, and Viet Nam have been involved or experience or at the minimum groundwork in both unified brackish water aquaculture-agriculture and secondly the aquaculture-silviculture.

Although during the recent times these sorts of experiences exist also in few of the more countries like the south of China and Malaysia.

Taiwan Province of China has proved to be the discoverer with reference to integrating marine seaweeds, specifically *Gracilaria* sp., in brackish water ponds with fish or shrimps (Figure 2.4) (de la Cruz, 1995).

Figure 2.4. Water salinity diagram.

Source: Image by Wikimedia commons.

In 2002, Linn stated in his overview which was based on "Aquaculture-aquaculture integration" put forward the classification: "Animals and

Animals" (fed fish/crustacean with filter-feeding fish and/or mollusks), "Animals and Plants" (fed fish/crustacean with macrophytes/seaweed), and "Animal and Plant plus Animal" (fed fish/crustacean with macrophytes/seaweed and grazing fish or mollusks).

He explained and gave examples, case studies and a more deep and complete theory of the three classes coming from both temperate and tropical regions. On the other hand, Chien, and Tsai (1985) classified pond farming into four categories, that is:

- Monoculture systems;
- Crop rotation culture systems;
- Polyculture systems (which culture several species at the same time in a single culture unit), and lastly; and
- Integrated culture.

Numerous species in separate units, which support and sustain exposure through the flow of food organisms and also of nutrients.

The outline of the very last classification (Chien and Tsai, 1985) the present review represents integrated mariculture systems into four major groups:

- Polyculture that deals with the multiple species co-cultured in a pond or tank or sometimes a cage (which also incorporates the inclusion of various species);
- Sequential integration (PAS: partitioned aquaculture systems) on land as well as in open water bodies (it is different from category a) by the requirement to channel a flow of wastes consequently in between the culture units with various species;
- Temporal integration (substitution of species enclosed in the same holding site, taking advantage from waste residuals that comes from preceding cultured species); and
- Mangrove integration (aqua silviculture, sequential practices-making the use of mangroves as biofilters).

There are various other instances when it comes to integrated systems, that are relevant under integrated mariculture practices, but these have been excluded in this survey. Out of the many, one example is the many forms of integration of brackish water aquaculture with agriculture as proposed by de la Cruz, in 1995 where animal wastes are put to use for the production of fish and shrimp (milkfish, tilapia, *Penaeus indicus*, *P. monodon*, etc.).

The sole integration with agriculture being incorporated is the rice-shrimp (*P. monodon*) farming. This happens due to such cultivation mostly occurs in regions with saline soils, establishes an important production, and incorporates a brackish water aquaculture species (while freshwater shrimps/fish are excluded from it).

The various other forms of integration that are not involved is because most of the time surround freshwater aquaculture includes: bacteria as biofilters, different halophytes, aquaponic systems, wetlands that is cattail and reed, however mangroves have been included), microalgae (along with the peculiarity of "green water" and occasional studies, *Macro brachium* spp. cultured in brackish water environments, species from similar feeding vocations (such as the mixes of two carnivorous fish), artemia cultivation, AquamatsTM, and building up of artificial reefs for the function of biofilter on a coastal scale.

Integrated mariculture in the tropics is in most cases discovered within a), c), and d), influenced by wider pond systems. More intensive technologies have been evolved and enlarged within b), particularly on land. Shrimp farming comes within all categories and identifies itself as the most studied system for integration under b).

This is not a big deal when taking into consideration the significance of shrimp farming in most of the tropical countries, and also because its connection with the deteriorating environment. Among many of the prevailing research projects (and theoretical abstract notions) based on integrated dimensions and perspectives that have been evaluated and discussed, only a small number out of them have been put to use and executed commercially.

2.5.2. Growth of Tropical Offshore Mariculture

During the current times, the mariculture farms within the tropical regions have experienced a gradual growth. Generally, the cage culture is going through a sudden growth and a deviation from simple, semi-intensive cultivation to the side of more intensive systems as stated by Halwart, Soto, and Arthur, in 2007, although this growth at times often ceases just short of the open ocean.

The development of offshore aquaculture takes shape to follow extensive trends of regional development, which signifies that the expansion of industry is not fundamentally restricted by the availability of sites, instead of the sites it is dependent on the accessibility of capital, technology, and expertise.

The most prosperous and flourishing aquaculture area to date has been Southeast Asia, especially China (Feng et al., 2005; Lovatelli et al., 2008; Halwart, Soto, and Arthur, 2007; Cao et al., 2007). This drift is astonishing in the dimension of the reduced availability of capital and the demanding and difficult hydrography of the enclosing seas (mainly storms), which prohibits the permission for the availability of technology somewhere away to be transferred, shifted, and implemented there without any delays or difficulties.

This was stated by Halwart, Soto, and Arthur, in the year 2007. Offshore farms have also been made and accepted in Hawaii (Marine Aquaculture Task Force, 2007), in the Caribbean (Bennetti et al., 2006, 2008) and in Oman (Al-Yahyai, 2008), and supplementary projects are organized and framed for tropical South and Central America (Bennetti et al., 2008; Stemler, 2009) as well as for Australia (Duckworth and Wolff, 2007).

Other tropical regions like African coasts, where the cage aquaculture is still trying to spread its wings and is a very new concept (Halwart and Moehl, 2004) lacks the offshore activity, apart from the European sponsored efforts in La Reunion, Mayotte, and Mauritius (Dabbadie, 2009).

The majority of growth has been in systems that are far away from shore, but stay in comparatively calm waters because of the fact that they are located in broad, shallow zones of the coastal shelf that frequently takes advantage from sheltering bays or in/near peninsulas.

This point outs the instabilities between the descriptive and functional definitions of offshore mariculture, in view of the fact that low-exposure offshore systems at times combine the advantages, difficulties, and needs of both coastal and open-ocean systems in different percentages and dimensions.

Explaining this further with an example, locating a farm which is three kilometers away the coastal area of precisely from shore in a sheltered bay would give rise to the high transportation costs also it advances and assists maximum automation whilst removing the requirement for heavy duty, storm-worthy net cages that are also submergible.

In total, the high cost and capital investment which is needed for offshore mariculture incites the cultivation of high-value species which is mainly for consumption in the variably rich markets. Though the carnivorous finfish grab the highest prices enclosed in the current global market, mainly the projects have focused on predator species which are precisely cobia, snapper, amberjack, seabream, red drum, pacific threadfin, and seabass.

This detaches itself from the majority of aquaculture in expanding and working for the development of tropical nations, which is dominated by cultures of shellfish, shrimp, and seaweed (that are kept for export) as well as milkfish and tilapia planned for the use and consumption in the regional and local areas (Troell et al., 2003). New species (that were previously uncultured) are, or in the recent past, being established and evolved for rearing in facilities provided in the offshore. These species incorporate tuna (Benetti and Watchinson, 2000; Kent, 2003; Ottolenghi, 2008), amberjack (Chen et al., 2008; Halwart, Soto, and Arthur, 2007), Pacific threadfin (Kam, Leung, and Ostrowski, 2003) and cobia (Liao et al., 2004; Bennetti et al., 2006, 2008; Nguyen et al., 2009).

Cobia culture has expanded from the beginning because of the exceptionally sudden growth rates (up to 6 kg/individual/year) and large market price of this species (Bennetti, 2006; Liao et al., 2004). Tuna, although highly profitable and renewable, is cultured as a value-added species only, though while talking about the life cycle of this species had not been closed and young tuna stocks must be captured in the wild (Ottolenghi, 2008).

A various number of projects in Spain, Japan, Croatia, and other nations are in the recent times developing and establishing brood stock and formulation facilities for tuna (Kent, 2003). Prior to this technology is, extensive aquaculture of tuna remains unfeasible and inaccessible, and one of the alternatives-harvesting of wild young tuna fishes-would result in the rapid depletion of wild stocks. Another important group of fishes that is regarded highly for offshore aquaculture is the flatfish.

Summer flounder, *Paralichthys dentatus* is popular and trendy in the United States of America, and few other countries as well. *P. dentatus* were the target of enormous research as the coastal and fishery managers tried their best to restock wild populations in Long Island Sound and to cater farm-reared fish in order to find the substitution of the wild caught fish when the flounder fishery plunged (Bengton, 1999).

Summer flounder were also brought up in offshore cages at Great Bay Aquafarms in Portsmouth, New Hampshire (USA). Flatfish like the flounder, sole, turbot, and halibut are very profit-making fishes in China, but their production was restricted due to their indoor cultivation which was done in ponds or tanks. The emergence of the Chinese submersible cages now permits farmers to rear and stern these flatfish in offshore farms (Chen et al., 2008) with a high-water quality and other growth and development conditions which provides benefits.

Moreover, a large number of projects have started experimenting with offshore-compatible systems (such as bottom cages, bottom lines, etc.), for growing shrimp, scallop, and mussels. These species may be are mainly useful as pollution-reducing complimentary species for finfish in offshore polyculture. Conchs that are grown near-shore in Turks and Caicos as well as Trochus in numerous Pacific-island nations could be enlarged offshore. Bottom sponge culture is a different branch of offshore mariculture which is being evaluated for productive and economic abilities to work successfully (Duckworth and Wolff, 2007).

Numerous finfish species have been proposed for the sustainable development of the future, including the yellowfin amberjack (*Seriola quinqueradiata*) and mahi-mahi (*Coryphaena hippurus*) (Abellán and Basurco, 1999).

2.5.3. Status of Offshore Aquaculture in Tropical Regions

The most initial difficulty and challenge which comprises while moving aquaculture offshore was in the context of technology which came with the urbanization and modernization. It proved to be a challenge because the conventional methods were appropriate for sheltered, and the environments that possess low energy.

Despite the destruction and damages done by the storms, the systems that were implemented and employed in the offshore have therefore by far did not fail the physical tests the ocean has put them through for examining, and must now experience other feasibility or viability issues.

In Southeast Asia, governmental subsidy, or moderation and entrepreneurship along with this the structural constraints have thrusted the offshore aquaculture to full commercial scale. China and Taiwan, Province of China have thousands of years of experience in terms of the aquaculture, thus the skilled labor is never difficult to find. In addition to this, producers have substantial biological and functional expertise regarding hatcheries, nurseries, feed, and grow out technology.

Consumers in these and other closest countries (mainly Japan and Republic of Korea) comprise of an almost demanding market for mariculture products. Moreover, new capital is immediately available and governments are keen to invest or finance the mariculture sector. There are national aquaculture schemes and plans in China and in Taiwan, Province of China to grow and flourish the sector and this is bound to assist in additional growth and development of offshore aquaculture in the near future.

This favorable "climate" is such that in these countries' user arguments or the criticism on environment is not probable to interrupt with offshore development. Constraints in Southeast Asia are most of the cases hydrographic and some researchers (Chen et al., 2007) assert that they will restrict production up to the considerable extent. Viet Nam is a new arrival in the industry of offshore aquaculture with credentials in the freshwater and inshore aquatic culture or segment.

Viet Nam also promotes a government-backed offshore plan and is viewed to become a prominent contributor to mariculture yields in the coming future. Other Southeast Asian countries who possess aquaculture experience are anticipated to progress offshore in the future as demands and prices keeps on escalating and given the risks are lowered, but during the recent years they are not developing in this direction.

The Caribbean region has its own different story where technology is analyzed in detail before licenses are given to anybody, governments subsidize and organizations invest. The climate is still one of the foremost discoverers and also the trial and error and therefore great plans are driven but have so far remained and endured mostly on paper.

This is principally true in the case of cobia, hailed as the significant culture species of the systems prevailing offshore. In terms of the long tropical coastline, Australia is another eminent country where the main cultured species is barramundi and some of the offshore farms that are still in the list of abandoned cases, instead of the large scale-up though it seems that the production will be delayed due to the legal and environmental constraints.

2.6. CONCLUSION

This chapter deals with the tropical mariculture and gives insight to the reader into the farming of fish in the tropics, the ocean fish farming in tropics and sub tropics. It tells the readers about the IMTA that is the IMTA in the marine temperate waters. It further explains the concept of IMTA and the current IMTA systems.

This chapter puts emphasis on the integrated tropical mariculture, and its systems classification, the growth of tropical offshore mariculture, and then towards the end it illustrates about the status of offshore aquaculture in tropical regions and in different countries.

REFERENCES

- Barrington, K., Thierry, C., Shawn, M. C., & Robinson, (2009). *Integrated Multi-Trophic Aquaculture (IMTA) in Marine Temperate Waters*. [online] ResearchGate. Available at: https://www.researchgate.net/publication/269996303_Integrated_multi-trophic_aquaculture_IMTA_in_marine_temperate_waters (accessed 10 September 2021).
- Dror, L. A., & Dor, E., (2013). *Sustainable Development of Marine Aquaculture Off-the Coast and Offshore: A Review of Environmental and Ecosystem Issues and Future Needs in Tropical Zones* (pp. 173–200). [eBook] Rome: FAO fisheries and aquaculture. Available at: http://www.fao.org/fishery/docs/CDrom/P24/master/root/11.pdf (accessed 10 September 2021).
- Geetha, S., & Viji, C. S., (n.d.). *Integrated Multi-Trophic Aquaculture Systems (IMTA)*. [eBook] CMFRI: Winter school on technological advances in mariculture for production enhancement and sustainability. Available at: http://eprints.cmfri.org.in/10666/1/7.%20Geetha%20Sasikumar.pdf (accessed 10 September 2021).
- Guerrero, R. D. III, (n.d.). *Fisheries and Aquaculture: Fish Farming in the Tropics*. [eBook] Encyclopedia of life support systems (EOLSS). Available at: https://www.eolss.net/sample-chapters/c10/E5-05-04-05.pdf (accessed 10 September 2021).
- Max, T., (2009). *Integrated Marine and Brackish Water Aquaculture in Tropical Regions: Research, Implementation, and Prospects* (pp. 47–131). [eBook] Rome: FAO fisheries and aquaculture technical paper. Available at: http://www.fao.org/3/i1092e/i1092e03a.pdf (accessed 10 September 2021).
- University of British Columbia, (2020). *Study: Ocean Fish Farming in Tropics and Sub-Tropics Most Impacted by Climate Change*. [online] Phys.org. Available at: https://phys.org/news/2020-02-ocean-fish-farming-tropics-sub-tropics.html (accessed 10 September 2021).

- Varsha, M. S., Joshi, K. K., & Thobias, P. A., (2015). *Winter School on Technological Advances in Mariculture for Production Enhancement and Sustainability*. [online] researchgate.net. Available at: https://www.researchgate.net/publication/308291220_Mariculture_and_Biodiversity (accessed 10 September 2021).

Chapter 3

Environmental Impacts of Mariculture

CONTENTS

This chapter deals with the environmental impacts of mariculture. It begins with explaining the ecological effects of bivalve mariculture, biological effects of mollusks which include biogeochemical cycling and habitat provision.

It goes on discussing various other aspects that is nutrient dynamics, biomineralization, habitat creation and maintenance. It then talks about the impacts of mariculture operations on ecosystems along with this it addresses the environmental impacts of shellfish aquaculture gear and practice.

This chapter also gives an insight to the reader about the impacts of mariculture on water quality which further sheds light on various dimensions such as the organic pollution, chemical pollution, eutrophication, antibiotics, etc.

Towards the end, it focuses on the impacts of mariculture on sediment along with explaining the releasement of substances having inhibitory effect on the aquatic environment.

3.1. INTRODUCTION

There is an imperative need to safeguard the ecology of nearshore marine habitats that are utilized for a steadily expanding scope of activities. Specifically, managers who are managing fisheries, need to contemplate both socioeconomic and environmental issues in coastal areas owing to the environmental changes that can happen because of the harvesting and cultivation process related to mariculture. Bivalve development can be extensively parted into three fundamental cycles:

- Collection of seed;
- Seed nursery and on-growing; and
- Harvesting.

The environmental impacts of each cultivation stage will fluctuate contingent upon the techniques utilized and the species in question. In numerous cases, prior to seeding, commercial species are reared as seed in hatcheries, with few consequences on the environment.

Although, while a few species are gathered from the wild utilizing benign techniques, for example, spat collectors, others are separated utilizing intrusive devices like dredges. An increasing number of studies on the ecological impacts of mechanical collecting devices have exhibited direct mortality of non-target species and the annihilation of appropriate settlement habitats or substrata.

Additionally, various other species, like starfish, birds, and crab might be denied significant habitat and food resources because of the mechanical harvesting of bivalve seed. The nursery and on growing of bivalves include either trestle culture intertidally, suspended culture sub tidally, or cultivation directly in/on the ground.

A large number of the environmental changes that happen result from their filter-feeding activities that produce pseudofeces and feces. This can lead to exhaustion of phytoplankton in densely cultivated systems and amassing of pseudofeces or silt underneath suspended cultures that then, many times result in a faunal impoverishment and anoxic environment.

In addition, environmental change can be caused by the structures utilized during the cultivation process. For instance, the utilization of netting to safeguard clams from crab predators prompts accumulations of sediment and siltation.

Parks of trestles can radically modify the water flow regime prompting changes in oxygen exchange and sedimentation rate within the framework. Broad intertidal cultivation plots could deny birds of feeding habitats and the related cultivation practices may upset perching birds. The last phase of cultivation includes harvesting (Figure 3.1).

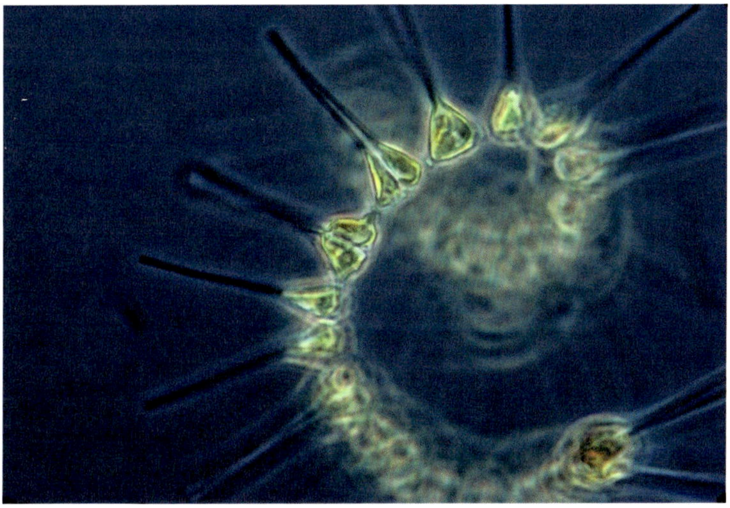

Figure 3.1. An illustration of phytoplankton.

Source: Image by Pixabay.

In numerous cases, this includes little more than unloading the bivalves from porches or lifting ropes. Be that as it may, on account of species relayed on the seabed or cultivated within the sediment, the utilization of intrusive techniques is required. The two suction and dredger gadgets cause turmoil of the sediment and straightforwardly eliminate or kill non-target species. A number of factors determine the time taken to recover by communities affected by these processes, like the aspect of the site and the strong characteristics of the sediment and the life span of the non-target fauna.

Just like the case with all anthropogenic exercises that influence the marine climate, the extent of the ecological changes that happen is connected to the size of the cultivation processes. There are likewise certain positive angles to coastal shellfish cultivation, for example, the arrangement of hard substrata and shelters in the sites which are otherwise barren and the likelihood of utilizing the cultured organisms as environmental sentinels.

Over the past 20 years, global bivalve aquaculture production has been increasing constantly. The primary bivalve species cultured on the planet are scallop, oysters, mussels, and clams. Since bivalve aquaculture requires minimal addition to the environment, it is considered to have a less dramatic environmental impact as compared to finfish culture.

> For the fundamental species cultured, the environment itself provides food and the ecosystem receives various nutrients and minerals from the waste. However, due to their effects on the suspended particles and their shell formations, the bivalves have the ability to create, maintain, and modify entire habitats.

Most of the research on the interaction between the environment and bivalve culture has been focused on oyster and mussel, presumably on the grounds that the production of various bivalves, for example, scallops, and clams is greater in Asia, where for centuries aquaculture and its effects has been a culturally accepted part of the coastal environment.

The objective behind this chapter is to introduce a view of the impacts of bivalve aquaculture on the surrounding environment, especially in coastal and estuarine zones, the current mitigation and prevention strategies and feature how bivalve culture can emphatically interrelate with the environment.

Bivalves are suspension feeders that play out their functions in a range of habitats, specifically lagoons, estuaries, and coastal oceanic systems. They acquire sustenance by filtering suspended particles, for example, detritus, and phytoplankton from the water segment. It has been determined that on average a shellfish can filter around 15 to 55 liters of seawater per day. By-products of bivalve are dissolved ammonium and bio-deposits of pseudofeces and feces. They confiscate nitrogen as protein in meat and shell and balance out phytoplankton growth dynamics through the moderation of ammonia cycling in the water segment.

They are along these lines considered 'keystone' species which apply 'top-down' control of phytoplankton by grazing yet in addition 'bottom-up' control through bio deposition and advancement of nutrient expulsion.

Epifaunal bivalves, for example, mussels, and oysters react to expanded levels of detritus and phytoplankton in the water segment with expanded filtration limit and production of pseudofeces. Infaunal bivalves, for example, clams instead of increasing the production of pseudofeces, adjust their clearance rates.

These cycles influence the food web, biogeochemical cycling, and the chemical and physical environment, conceivably modifying habitats and ecological functioning. Therefore, bivalve aquaculture has the capacity to influence the climate in both positive and negative manners, with an assortment of all over field cascading effects on various parts of the ecosystem, influencing secondary and primary productivity and community structure.

Sedimentation rate, water flow and sediment composition can be altered by culture structures and operations and it can also disturb the benthic flora and other marine organisms. Estuaries are frequently a favored site for bivalve culture and in this manner are the environments analyzed by most of the literature.

Oysters and mussels maintain a high clearance rate and reject large numbers of particles as pseudofeces and thus they exert the highest influence on the environment. The immediate correlation of impact between mussel (*Mytilus galloprovincialis*) and clam (*Tapes philippinarum*) culture situated in a similar water body has indeed shown a more prominent impact for mussel culture.

All things considered; these three cycles can happen simultaneously. In light of Gavine and McKinnon's danger evaluation for blue mussel (Mytilus edulis) and oyster (Pacific, Sydney rock and pearl) culture in Australia, the higher danger of effect was related with the alteration of sediment physical structure, the crumbling of sediment quality, and the effect on seagrass beds.

The impacts of bivalve aquaculture on the environment have been classified in various manners all through the literature. For the purpose of this manuscript, they will be split into four primary impacts: on the sediment, on the water segment, impacts of bivalve introduction and consequences for other marine species (including marine birds and mammals).

3.2. ECOLOGICAL EFFECTS OF BIVALVE MARICULTURE

Through research in biogeochemistry, fisheries science, ecology, interdisciplinary marine science, physiology, and biogeochemistry, the role of suspension feeding bivalves in estuarine and marine ecosystems has been widely documented. Suspension-feeding bivalve mollusks devour at the lowest trophic level, feeding largely as herbivores (Duarte et al., 2009). There are three areas to describe:

- The biological activities of mollusks (regardless of whether wild or refined) and the impacts of their habitat provision and biogeochemical changes;
- The incidental effects of bivalve mariculture procedure on numerous segments of the ecosystem brought about by mariculture structures and activities and by the biological activities of the mollusks; and
- Outcomes of activities taken by culturists to adjust ecological interactions deliberately to deal with the impacts of competitors, predators, and pests on mariculture systems.

3.2.1. Biological Effects of Mollusks: Biogeochemical Cycling and Habitat Provision

Benthic suspension feeders, like numerous types of bivalve mollusks, impact the nutrient and organic coupling of pelagic and benthic (Dame, 1996) through their capacity to filter a wide size range of particles and store organic wastes that sink to the base (bio deposition).

Suspension-feeding bivalves play out this function in a range of habitats and physiographic conditions (e.g., Coastal oceanic systems, lagoons, estuaries) where they filter out and store notable measures of suspended material, just as excrete dissolved nutrients.

In estuarine frameworks, the impact of benthic suspension-feeding bivalves on turbidity, habitat complexity, benthic-pelagic coupling, deposition, nutrient remineralization, and primary production has been very much archived (Dame and Olenin, 2005).

Kaiser (2001) evaluates the impacts of molluscan cultivation on the ecology of systems, distinguishing a comparable arrangement of mechanisms of influence, and reasons that such Processes generally have a positive influence on the overall water quality of a system, which is very well depicted for intertidal oysters by Dame (2005).

3.2.2. Nutrient Dynamics

Through direct excretion, mollusks impact nutrient dynamics and it also influences indirectly through microbially mediated remineralization of their organic deposits in the sediments (McKindsey et al., 2006a). Accordingly, nutrient regeneration is identified with the location and abundance (shallow versus profound water) of bivalves in a framework.

The degree to which these influences overall nutrient budgets and along these lines primary production is related to the system flushing rate and residence time (Dame, 1996; Newell et al., 2005). The ensuing extents of components in the framework will impact the degrees of recycling and possibly bring about at least one or more being limited (Dame, 1996).

Since nitrogen is the most common nutrient-limiting biological production in marine and estuarine systems, most of the studies of bivalve impacts on nutrient recycling have focused on it. (Parsons et al., 1983; Howarth, 1988; National Research Council, 2000).

Benthic bivalves are significant contributors of nitrogen (normally as ammonium, NH_4^+) to both intertidal and subtidal frameworks. Nixon et al. (1976) infer that nitrogen is heavily influenced by tidal flow and its flux across oyster reefs is highly variable.

Dame (1986) evaluates a collection of work identifying nutrient fluxes including *Crassostrea gigas* in northern France and infers that 15 to 40% of the nitrogen in the system was obtained from the oysters. Moreover, estimated values were consistently higher than the assessed values, likely because of remineralization happening in adjacent sediments.

The idea that the macroalgal front of mussel beds will seize nitrogen (Asmus and Asmus, 1991), accordingly making it inaccessible for phytoplankton production, has likewise been proposed for other systems (Mazouni et al., 1998; Lin et al., 2005).

Conversely, nitrogen is perpetuated within certain frameworks through direct recycling of nitrogen from bivalves (e.g., *Crassostrea virginica*) to phytoplankton (Dame and Libes, 1993; Newell et al., 2005).

Various studies have shown that nutrients obtained from bio deposits or potentially discharged nitrogen serve to upgrade the development of eelgrass and other submerged aquatic vegetation (see underneath).

In the Marennes-Oléron culture locale in France, Leguerrier et al. (2004) demonstrate that higher production of oyster expanded benthic-pelagic coupling, which thus expanded secondary production (in the form of meiofauna), giving food to adolescents of predacious nektonic species (Figure 3.2).

Figure 3.2. An illustration of mollusk.

Source: Image by Pxhere.

Likewise, Mazouni (2004); and Newell et al. (2005) exhibit that other planktonic organisms (flagellates, bacteria, and ciliates) can go about as wellsprings of nitrogen for bivalve mollusks when suitable autotrophic phytoplankton is not present.

Phosphorus is essential to biological systems, and phosphorus budgets constructed in and around mollusk gatherings show extensive expulsion of this supplement from the system through bio deposition. Asmus et al. (1990) show that mussel beds with enormous macroalgal populaces delivered less phosphate than beds without a huge macroalgal component.

Silicon is a significant component for diatoms and can be restricted in systems dominated by diatoms. Bivalve mollusks add to the recycling of silicate through shifting of this nutrient from the water segment to the sediment with little being sourced from the bivalves (Prins and Smaal, 1994).

Mollusks, like mussels, may likewise specifically sustain on segments of particulate matter and accordingly concentrate certain metals like copper in their pseudofeces (Allison et al., 1998).

The production of pseudofeces in huge amounts is a significant mechanism by which bivalves couple the water segment to the base (Dame, 1996). Epifaunal bivalves (mussel and oysters) have a plastic reaction to expanding levels of detritus and plankton in the water column with steadily expanding production of pseudofeces and filtration capacities.

Although, this response is not seen in infaunal bivalves (cockles and clams), which direct ingestion rates at high-seston concentrations by changing clearance rates as opposed to by expanding production of pseudofeces (e.g., Foster-Smith, 1975; Bricelj and Malouf, 1984; Bricelj et al., 1984; Prins et al., 1991; Iglesias et al., 1996).

Mussels and oysters are likewise known to endure moderately significant levels of suspended inorganic particles and proceed to filter and produce higher levels of bio deposits.

The negative and positive feedback mechanisms perceived in aquatic systems as a result of nutrient dynamics intervened by mollusks have been the subject of various studies (Dame, 1996; Prins et al., 1998; Newell et al., 2005).

Their rapid response to high levels of food (e.g., plankton), relative permanence in aquatic systems and high filtration capacity, enable bivalves to balance out frameworks and improve flexibility to perturbations (Jackson et al., 2001a; Newell, 2004).

Enormous bivalve accumulation can control the abundance of phytoplankton in shallow seas (Newell et al., 2005; McKindsey et al., 2006a) for a rundown of applicable studies), and extreme filtering can decrease phytoplankton bloom intensity while broadening the length of less intense blooms (Herman and Scholten, 1990).

Bio deposition and filtration of phytoplankton and other suspended materials by large beds of bivalves likewise diminish downstream transport, in this way decreasing impacts of excess nutrients or sedimentation in outlying waters.

Hence, bivalves give the system an ability to support against abrupt perturbations (DeAngelis et al., 1986; Jackson et al., 2001a; Lotze et al., 2006). The large-scale expulsion of bivalves from a system has brought about some very much reported changes in system processes and has added to the general debasement of water quality or, all the more suitably, a decrease in the flexibility of the system to perturbations like supplement sedimentation and nutrient loading (e.g., clams in the Chesapeake Bay; Newell et al., 2005, 2007; Pomeroy et al., 2006).

Numerous estuaries, like the Chesapeake Bay, and coastal oceans experience the ill effects of eutrophication, where surplus nutrients enter streams from atmospheric deposition and land-based sources (e.g., industrial releases of nitrogen oxides or ammonia, sewage treatment plants, agricultural use of fertilizers and farm animal wastes) and trigger huge blooms of phytoplankton and other green growth.

Phytoplankton blooms lessen water clearness and exhaust the water of oxygen as they die and decompose. Bivalves can diminish exorbitant growth of phytoplankton and, at high density, can balance symptoms of eutrophication, along these lines improving local, and sometimes downstream, water quality.

However numerous bivalve mollusks have been drained by overfishing, particularly oysters (Jackson et al., 2001a; Kirby, 2004; Lotze et al., 2006; Beck et al., 2009), and in addition clams (Peterson, 2002; Kraeuter et al., 2008) and scallops (Peterson et al., 2008). Subsequently, increasing suspension-feeding bivalves, ideally local, through mariculture and restoration can possibly upgrade suspension-feeding activity and controls in systems where common populaces have been exhausted (Jackson et al., 2001a).

3.3. BIOMINERALIZATION

Additionally, nutrient cycling, mollusks add to biogeochemical measures through shell development, which catches carbon as calcium carbonate and can lead to sequestration of carbon in marine sediments after regular mortality of wild mollusks or earthly entombment of shells after consumption of wild-caught or cultured mollusks.

The shells of mollusks (living and dead) aggregate in different kinds of constructions in coastal, estuarine, and oceanic systems (Shumway and Kraeuter, 2000). Surface shell collections give a range of ecosystem services, essential among which are structural habitat (e.g., shelter, intricacy) and erosion reduction (Coen and Grizzle, 2007).

Shell is likewise a significant wellspring of sedimentary carbonate content. In today's time the carbonate budget of estuarine and coastal water has become a cause of concern because of the extensive shell extraction (harvesting and mining for construction), the restrictive expense of long-haul continual substrate provisioning to help fisheries, and the deficiency of shell by means of diminished bivalve populaces outcoming because of fishing and infection measures (Mann and Powell, 2007).

Also, increasing ocean acidification brought about by expanding concentrations of atmospheric CO_2 has genuine ramifications for seawater carbonate chemistry (Brewer, 1997; Caldeira and Wickett, 2003; Feely et al., 2004; Doney et al., 2009).

Recent studies have demonstrated that bivalve development, survival, and growth are adversely affected by diminished pH (e.g., Berge et al., 2006; Fabry et al., 2008; Kurihara, 2008). The adjustment in carbonate water chemistry and collateral reduction in the feasibility of bivalve mollusks conceivably will lessen both the persistence and provisioning of shells in estuarine and coastal systems, especially those in high latitude regions with low alkalinity seawater (Feely et al., 2004; Lee et al., 2006). Accessibility of plentiful mollusk shells in the surface dregs can give local buffering against increasing acidity.

The significance of the interconnection between sedimentary carbonate content and ecological communities was segmented in an applied model that depicted a positive input process between benthic mollusks and carbonate addition to the sediments.

The taphonomic (cycle of fossilization) input theory hypothesis underlying this conceptual model (Kidwell and Jablonski, 1983) asserts that expanding shell content supports the settlement of calcifying organisms, and their demise increases the pace of carbonate addition, shaping a process for the positive feedback.

In recent studies, it is stated that the interactions between community dynamics and carbonate content are crucial to ecosystem dynamics in estuarine systems (Gutierrez et al., 2003; Powell et al., 2006; Powell and Klinck, 2007). The species profiting most are the carbonate makers, especially bivalves that, through their own demises, give a crucial sedimentary constituent advancing the long-term endurance of their species.

Shell is a fundamental part of present-day estuarine and coastal ecosystems, although it is not a steady asset (Powell et al., 2006). The shell should be constantly renewed and will vanish quickly if the cycles that help this restoration are eased back or halted. Carbonate loss potentially surpasses gain in shallow-water marine ecosystems today (see the conversation in Powell et al., 2006). All things considered, commercial mariculture practices and current environmental conditions, when combined with anticipated changes, like sea fermentation, will accelerate and facilitate carbonate loss in coastal and estuarine systems. In this way, the management of shell-producing commercial species should likewise incorporate the administration of the natural surroundings that will boost the production of carbonate.

Long haul sustainability of mollusk stocks relies on the upkeep of a positive shell budget for carbonate, just as the arrangement of natural surroundings that upholds survival, recruitment, and growth of bivalves. Mariculture of bivalve mollusks can contribute well to the production and preservation of shell in coastal ecosystems if after harvesting the operators return back the resources of the shell to the environment. Nonetheless, to achieve this goal, guidelines requiring the return of shells to the coastal bottom, estuarine, or lagoonal, after sucking may be required.

3.4. HABITAT CREATION AND MAINTENANCE

Shell adds hard substrate and natural surroundings intricacy to delicate substrates, subsequently expanding species variety (Wells, 1961; Larsen, 1985; Coen et al., 1999; Harding and Mann, 2000, 2001; Mann, 2000; Gutierrez et al., 2003) and improving survival and recruitment of bivalves (Haven and Whitcomb, 1983; Abbe, 1988; Kraeuter et al., 2003; Bushek et al., 2004; Green et al., 2004; Soniat and Burton, 2005).

Shell adds bottom habitat complexity to the ecosystem when they are present in significant amounts (Haven and Whitcomb, 1983; DeAlteris, 1988; Grizzle, 1990; Powell et al., 1995; Allen et al., 2005). Fish have been appeared to connect with both artificial and biogenic structures on the base, like the legs of oil platforms, eelgrass, and bivalve reefs, as a result of appreciation for organized natural surroundings for feeding and protection (Franks, 2000; Heck et al., 2003; Peterson et al., 2003; Coen and Grizzle, 2007; Horinouchi, 2007; Jablonski, 2008).

Seagrasses are usually viewed as a very significant plant in lagoons and estuaries where they construct a potential structural habitat for invertebrates and fish in these soft-sediment systems (Jackson et al., 2001b; Williams and Heck, 2001; Heck et al., 2003; Bostrom et al., 2006). Local enhancements in water lucidity initiated by filter-feeding bivalves can advance the spread of eelgrass, particularly to profundities where light would some way or another be restricting (Dennison et al., 1993).

Eelgrass growth can also be stimulated by augmentation of nutrient concentrations in sediments, as has been appeared to happen for eelgrass growing close by mussels in southern California, Europe, and Florida (Reusch et al., 1994, Reusch and Williams, 1998; Peterson and Heck, 1999, 2001a, b).

Numerous estuaries on the west shoreline of the United States are flushed with relatively supplement-rich ocean waters, and under these conditions, eelgrass may not benefit as much from the extra supplements liberated by bivalves (Figure 3.3) (Dumbauld et al., 2009).

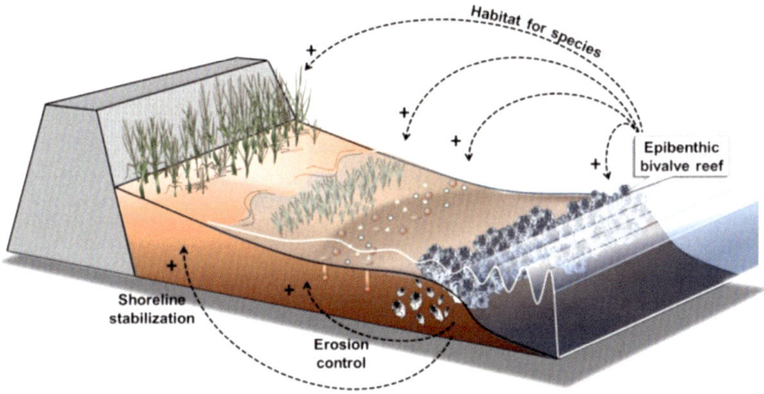

Figure 3.3. Ecosystem services delivered by epibenthic bivalve reefs.

Source: Image by Wikimedia commons.

Both the decrease of fertilizing and turbidity impacts of bivalve mollusks have been exhibited tentatively for modest densities (16 for each m^2) of hard mollusks (mercenaria) in a moderately oligotrophic Long Island estuary (Carroll et al., 2008).

Beneficial outcomes of a humble number of suspension-feeding bivalves are bound to profit eelgrass in generally oligotrophic water bodies, where without a huge increase in filtering capacity, functional enhancement of water clarity may be achieved (Carroll et al., 2008).

3.5. IMPACTS OF MARICULTURE OPERATIONS ON ECOSYSTEMS

Mariculture can have both adverse and positive impacts upon populations of huge marine vertebrates, like marine turtles, birds, and marine mammals.

Practically all studies on these communications have zeroed in on finfish farming and the ecological and economic effects that outcome from depredation (e.g., Nash et al., 2000). Of the limited study on the effect of bivalve mariculture on untamed life populaces, most related to bird populaces.

Three studies which were published earlier investigate effects on marine mammals (Markowitz et al., 2004; Watson-Capps and Mann, 2005; Becker et al., 2009), yet they were not planned explicitly to identify ecological effects on these species, and just Becker et al. (2009) relate to bivalve mariculture techniques presently utilized in the United States. There are no published studies on potential interactions with estuarine or marine turtles.

Drawing upon a more extensive comprehension of the ecology of these species, likely effects of bivalve mariculture upon these wildlife populaces have been recognized in one published review (Kemper et al., 2003).

Also, a National Oceanic and Atmospheric Administration (NOAA) workshop (Moore and Wieting, 1999) investigated more extensive associations among mariculture and marine turtles and marine mammals, and a conversation paper on the likely impacts of mussel farming on seabirds and marine mammals was delivered by the New Zealand Department of Conservation (Lloyd, 2003). It ought to be noticed that immediate display of these effects is uncommon, and in most of cases, potential impacts are in this manner anticipated from the best existing data.

A significant reason for mortality for marine mammals, seabirds, and estuarine and marine turtles is entanglement in fishing gear and marine debris (Lewison et al., 2004; Read et al., 2006). Trap in mariculture gear is very uncommon, yet two Bryde's whales have purportedly died in independent events after entanglement in mussel spat assortment ropes in New Zealand (Lloyd, 2003), and marine turtles have additionally been trapped in ropes (Godley et al., 1998; Kemper et al., 2003).

The presentation of any netting or lines, for instance, to reject hunters, may along these lines represent a danger of entanglement to marine mammals, birds, and estuarine and marine turtles. Where bivalve mariculture activities venture into offshore areas, this may improve the probability of associations with sea turtles and enormous whales, secured in the United States under the Endangered Species Act. Based on experience with different kinds of mariculture and fishing activities, the dangers of traps can be decreased by utilizing heavier lines and guaranteeing that anti-predator nets and lines are kept tight.

Ingestion of marine litter is likewise known to cause mortality in marine mammals, marine turtles, and birds, (Derraik, 2002). Mariculture tasks are perceived as a significant origin of marine litter (Johnson, 2008). For instance, youthful Australian gannets in New Zealand's Marlborough Sound have been discovered entrapped in rope ties from mussel farms that have been integrated into their nests (Lloyd, 2003).

In several ways, mariculture activities may likewise impact prey accessibility for marine mammals, birds, and estuarine and marine turtles. As examined before, the presence of culture sacks will change the design of benthic communities and the degree of eelgrass beds, incidentally influencing prey availability.

Additionally, Manila clam cultivation in sacks adversely influences the utilization of supported foraging regions by oystercatchers (Godet et al., 2009). Then again, farm structures may expand food accessibility by giving a substrate to biofouling organisms acceptable as prey, like mussels.

In British Columbia, for instance, densities of Barrow's Goldeneye (*Bucephala islandica*) and Surf Scoter (*Melanitta perspicillata*) and were emphatically connected with the existence of oyster farms (Žydelis et al., 2009).

Since these species do not eat oysters, this connection seems to be driven by the high densities of mussels recorded on mariculture structures (Kirk et al., 2007). For this situation, predation by sea ducks on wild mussels was not seen as negative by shellfish ranchers (Žydelis et al., 2009).

Although it has been shown that predation by eiders can reduce muscle biomass by 50% in various experiments on natural mussel beds (Hamilton, 2000), exhibiting the effect that sea ducks can have upon commercial mussel farms in certain regions.

In intertidal regions, the presence of culture bags may straightforwardly disbar shorebirds that explore in the sediment from foraging habitat (Kelly et al., 1996). Würsig and Gailey (2002) likewise feature the need to think about a possible loss of breeding and feeding habitats for cetaceans because of the actual presence of bivalve mollusk farms, especially given anticipated expansions in these facilities in inshore environments.

Succeeding researches in New Zealand and Western Australia (Markowitz et al., 2004; Watson-Capps and Mann, 2005) demonstrate that dusky dolphins (*Lagenorhynchus obscurus*) and bottlenose dolphins (*Tursiops spp.*) stay away from farmed regions where mussel and oysters cultivation hanging lines are available.

Deracination from key regions may likewise result from disruptions owing to the activities of mariculture laborers (Becker et al., 2009). This disruption might be caused by the presence of laborers in intertidal regions or by boats related to mariculture activity. Likewise, marine mammals may react to commotion from mariculture-related boat traffic.

Mariculture constructions can give shelter, haul-out sites, or roosts for seals and birds. This is probably not going to adversely impact seal or bird populaces, yet it might improve the probability that these species cause fecal defilement of mollusk beds.

It has likewise been noticed that the presence of mariculture constructions could draw in adolescent marine turtles, which typically pile under patches of hovering weed, subsequently upsetting natural dispersal behavior (NOAA, 1999).

Data on the possible impacts of mariculture laid out above is to a great extent dependent on an overall comprehension of wildlife ecology and the connections of these species to the biological and physical environment instead of dependent on directed studies built around mariculture activities.

Indeed, even where studies have been done around shellfish ranches, lack of certainty over spatial and temporal variation in both the location of structures (Watson-Capps and Mann, 2005) and levels of disruptions (Becker et al., 2009) oblige the conjectures that can be drawn about the effects of mariculture.

Although, there is less uncertainty about the overall impacts of 'disruption.' The tending of any mariculture activity requires a presence of humans, and numerous researches have utilized evasion distances to set up support zones to limit disruption from other human activities (e.g., Rodgers and Smith, 1997; Blumstein et al., 2003).

However, it is significant to acknowledge that few species might not show avoidance if there is an insufficiency of suitable alternative habitat, even where the wellness costs are high, and disruption expenses may along these lines be miscalculated or undetected (Gill et al., 2001). Thus, evaluating whether disruption has a populace consequence, assessed as expanded mortality or diminished fertility, is a considerably more troublesome proposition (Stillman et al., 2007). Likewise, insubstantial comprehension of the foraging distribution of marine turtles, birds, and marine mammals from spatially restrained rearing colonies additionally makes it immensely challenging to evaluate populace level effects of entanglement, disturbance, or loss of habitat happening from bivalve mariculture. Consequently, if there is expanded mortality around a culture site, there could be ramifications for rearing colonies many kilometers away.

3.6. ENVIRONMENTAL IMPACTS OF SHELLFISH AQUACULTURE GEAR AND PRACTICES

In an evaluation of the environmental effects of shellfish aquaculture, Kaiser et al. (1998) differentiate among the effects of cultural organisms and the practices needed for harvesting and growing them. For example, in Southern New England and some parts of the Northeast, it is a standard practice to rear oysters in a subtidal rack and bag system (Figure 3.4).

Figure 3.4. An illustration of oyster farm shellfish.

Source: Image by Pixabay.

Evaluation of the effect of this gear type proposes that they go about as refugia for a variety of marine organisms, including the adolescent phases of different types of commercial valuable finfish (DeAlteris et al., 2004; Tallman and Forrester, 2007).

There is additional proof that shellfish in suspended culture enhance fish and crab populaces on the base (Iglesias, 1981; Mattson and Linden, 1983), and that fouling organisms on the mussel long lines can upgrade populaces of predatory and grazing fish (Tenore and Gonzalez, 1976). In certain regions of the Northeast, soft shell clams and quahog clams are farmed in intertidal flats where plastic mesh or nets are utilized for covering the shellfish and protecting against hunters. Investigations of effects of this sort of shellfish farming in England utilizing the Manila clam Tapes philippinarum have exhibited buildup of fouling organisms on the netting and an accompanying expansion in grazing mollusks and adolescent fish related with the nets (Spencer et al., 1996).

While the arrangement of the beds and gathering of clams from the plots by different hydraulic devices or mechanical plows disturbs infaunal communities, recuperation is fast with original communities returning in under a year (Spencer et al., 1998).

Disruption and inadvertent take of non-target organisms are unavoidable despite harvest mechanisms and bed readiness; although, in most cases increased diversity and plenty of species living in little hiding spots of all shellfish make up for temporary losses (Kaiser et al., 1996).

3.7. IMPACTS OF MARICULTURE ON WATER QUALITY

Impacts of waste on water bodies rely upon the critical attributes of the waste quantity, habitat, time-scale over which the releases happen and attributes of the water column. The wastes originated from aquaculture can be subdivided into two major groups; 1) soluble waste, 2) solid waste.

The solid waste consists of feces, mucus, and uneaten pellet feed. Since the majority of them are heavier than water they sink to the base. Soluble wastes are Urea, ammonium, and some particles that are dissolved from solid waste. The main components that affect water quality from these wastes are nitrogen, phosphate groups and suspended substances.

In addition, a portion of the minerals and vitamins (minor components) are effective on algae populaces. Pollution originated from aquaculture is generally divided into two categories; 1) organic pollution 2) chemical pollution.

3.7.1. Organic Pollution

The primary sources of organic pollutants are dead individuals, fish feces and uneaten feed. Feces excreted by fishes are connected with the ingredients of the feed. Around 25% to 50% of the feeds consumed are defecated to the environment in the form of feces. Intensive fish farming causes a lot of organic waste deposition in the water column and sediments (Figure 3.5).

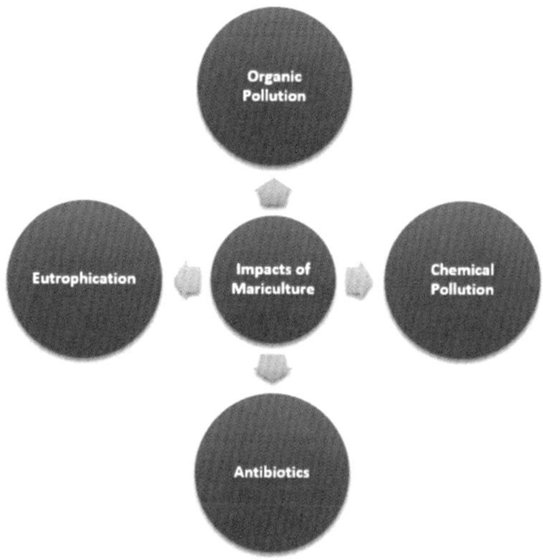

Figure 3.5. Impacts of mariculture on water quality.

A sediment gatherer put under the cages for salmon cultivation gathers 14.7 kg to 52 kg of organic waste in each square meter per annum, which uncovers the degree of organic pollution emerging from aquaculture activities. Several ecological issues arise due to aquaculture activities.

For example, aquaculture facilities, like cage systems, remarkably lower the speed of the water current hence cause deposition of organic pollutants in the surrounding environment. Due to the decomposition of these organic waste substances, the quality of water deteriorates and the requirement for water oxygen increases.

Adverse impacts like accumulation of organic waste in the sediment, accumulation of toxic substances, turbidity, formation of anoxic zones near the seafloor, and spread of diseases are prominent in areas where aquaculture is practiced, particularly in restricted exchange environments.

The formation of various gasses like CO_2 (Carbon dioxide), H_2S (hydrogen sulfide) and CH_4 (Methane) on sediment slows down the growth rate of living organisms in the local region and makes them more assailable to diseases.

Decomposition of organic waste originating from aquaculture activities directly reduces the level of oxygen in the water. Due to the release of nutrients like phosphate and nitrogen into the water column, the quality of water changes.

Excessive amounts of organic supplements boost the growth of algae and water plants. At the point when algae that produce toxins grow excessively, they can reach blooming concentrations and cause adverse effects to various organisms.

Cultured fish may die off because of oxygen deficiency and disorder occurring in the gills because of changes in water quality associated with the abrupt growth of algae. Individuals who consume biotoxin contaminated shellfish can get poisoned and may die as well.

When algae die in huge numbers, dead organic substances begin to decompose and, in this process, a lot of oxygen is consumed which leads to hypoxia. Deposition of inorganic and organic wastes in regions with low water exchange rates can lead to hyper-nitrification as well.

In this sense, in such regions, permanent aquaculture activities accelerate the accumulation of waste. On the other hand, aquaculture activities carried out in regions with a high bottom current rate or in deep waters may result in widespread wastes.

To explain the extent of pollution caused by aquaculture activities, let us make a comparison. In a year, an average person excretes 1.1 kg of phosphate and 4 kg of nitrogen, on the other hand in a year every ton of salmon excretes 4.8 kg of phosphate and 55 kg of nitrogen into the water column through feces.

A salmon cultivation facility with a production capacity of 50,000 tons per year releases an amount of phosphate equals to that of 216,000 people into the water in a year and an amount of nitrogen equal to that of 682,000 people.

3.7.2. Chemical Pollution

An assortment of chemicals utilized for disease outbreak treatment and disinfection in aquaculture. Environmental effects of chemical compounds rely upon pathogens' sensitivity to antibiotics, the poisonous impact of the compound, therapeutics, and active time zone of the chemical compound in the environment.

The majority of the chemicals were modified from cattle husbandry and poultry and their viable impact on the marine environment is not fully researched. Chemicals utilized in mariculture can be further divided into antifoulings, veterinary medicines and disinfectants. Principle veterinary medicines pesticides, antibiotics, and anesthetics.

Although veterinary medicines are not so common in bivalve cultures and crustaceans but are widely used in fish farms. Majority of the chemical treatment scenarios in mariculture end in the release of chemicals directly into sediments and water columns. For the cage systems rainbow trout, sea bream, sea bass, Atlantic salmon, are the most well-known fish species and the vast majority of the chemicals utilized for which cause pollution.

Among various meanings of marine pollution, the most well-known one is GESAMP's (joint group of experts on the scientific aspects of marine pollution), which states that marine pollution is "the damage to living creatures in the environment, the hindrance to marine activities including fishery, the inability to use seawater and the reduction of amenities as a result of changes in environmental conditions due to deleterious effects induced by various waste materials and/or energy left to the marine environment (including estuaries) by people directly or indirectly."

Farmers try to control the outbreak of the diseases by utilizing antibiotics. As a consequence of this, antibiotics make their way through water by uneaten antibiotic-treated feed or feces. Different kinds of pesticides are used in order to control external parasites such as sea lice. Pesticides could be given to fish via means of bath or feed.

Pesticides, for example, ivermectin have high poisonous impacts on marine organisms. One of the serious issues in cage farming is biofouling. The proliferation of bio-fouling organisms adds a substantial load to the open water cage net which dangers the steadiness of the structure and promptly reduces water flow which deteriorates the quality of water within the cage systems. Anti-fouling coatings and paints are utilized to moderate the growth of such organisms on the net to keep up the water flow through the systems. Generally, copper is the active matter of such paints.

With time decay of paint causes copper to infiltrate water, which presents poisonous impacts on marine animals. Another poisonous metal that originated from aquaculture is zinc. Zinc sulfate ($ZnSO_4$) is utilized as a feed supplement to avert the outbreak of diseases.

3.7.3. Eutrophication

Eutrophication basically implies improved sustenance and alludes to the incitement of phytoplankton or macro fit (plants and algae) growth in the aquatic environment. Until now, in all modeling researches, just impacts of mariculture originated from nutrient waste on phytoplankton development were researched. Particularly, phytoplankton growth in warm months constrained by nutritious components (phosphate, nitrogen, and so on) was demonstrated.

Variables of this sort of model may line up as; water column affected from waste and dilution rate, calculation of nutrient level, determination of the relationship between phytoplankton growth and nutrients and determination of bottom structure.

In the Baltic Sea, connections between seawater sensitivity, (relies upon the interrelation between coastal water and open water) concentration of nutrients, phytoplankton biomass and Secchi profundity contrasted by utilizing various regressions techniques. Contingent upon the sensitivity of the coastal regions, Secchi profundity decreases, while phytoplankton biomass increments.

However, since the water in shallow regions is under the impact of wave-induced re-suspension, Secchi's profundity strategy is undependable. In this manner, the most precise technique for the determination of the phytoplankton biomass is the estimation of chlorophyll content.

In another research, the indirect and direct impact of eutrophication was researched between the years 2001 to 2011 in the Baltic Sea by utilizing Secchi profundity markers, nutrient concentration, and chlorophyll-a. The study revealed that between 2007 and 2011, the whole open Baltic Sea was affected by eutrophication.

3.7.4. Antibiotics

In fish health management, chemicals have a wide range of usage. Reports prepared by WHO (World Health Organization) and mention use of chemicals represents a danger to human wellbeing. In spite of the decrease in organophosphate and antibiotic use in aquaculture, colorant, anti-parasitic, synthetic parathyroid, preservatives, and various other chemicals that cause marine pollution are still used.

The utilization of chemicals influences the marine environment as well as represents a danger to the farmers also. One of the major problems in aquaculture is the random utilization of antibiotics for fish health management.

Unmetabolized antibiotics are regularly passed to the aquatic environment by feces and uneaten feeds treated with antibiotics. In both sediments and water columns, various classes of antibiotics have diverse half-lives.

Oxytetracycline which is perhaps the most widely recognized antibiotic can be efficacious as long as 30 days in both sediment and water. Indeed, oxytetracycline can be measured on sediments even after many months without being active biologically.

In cage systems, chemicals in feces and uneaten feeds treated with antibiotics disperse in the surrounding environment by current and waves. Filter feeders and wild fish, for example, oysters, and mussels devour uneaten feeds treated with antibiotics or it passes to sediment.

In aquaculture, bioactive substances, for example, pesticides, and antibiotics are mostly utilized for controlling parasites and diseases in aquaculture. Achievement or disappointment in aquaculture relies upon the right utilization of chemicals, duration of usage conditions in which it is used, against parasites and infectious diseases. Impacts of bioactive matters on the environment can be aligned as:

- In animal tissue substances with inhibitory effects last longer;
- Unleash of substances with inhibitory effect to the aquatic environment;
- Transmission and development of the antibiotic resistance gene in microorganisms.

3.8. IMPACTS OF MARICULTURE ON SEDIMENT

Till date, the significance given to the investigation of the effects of synthetic paints, pesticides, and antibiotics on the health of the human has not been given to the studies related to investigating the effects of these chemicals on the marine environment. After feeding fish in a cage, feces, and uneaten feed accumulate on the seafloor.

Inordinate gathering of organic substances on the seafloor leads to excessive multiplication of microorganisms which causes a decrease in levels of oxygen needed for other benthic living beings (fish, crustaceans, and so on).

The outbreak of diseases happens once in a while in wild fish populaces since they are dispersed everywhere in the seas and oceans, though high populace density in aquaculture systems frequently prompts the outbreak of disease. Amassing of antibiotics, uneaten feed, chemicals, and feces cause the development of an anoxic zone on the seafloor in this manner, the majority of the benthic organisms that live close to the salmon homesteads disappeared, just a portion of the resistant bacteria multiplied. In numerous nations, the development of antibiotic resistance to antibiotics in cattle, poultry, and aquaculture-borne bacteria threatens human wellbeing.

Antibiotics utilized against outbreaks of bacterial fish disease are for the most part discharged to the water column. These antibiotics amass on animals or sediment in the surrounding environments. Researchers identified nine distinctive antibiotic compounds in various stream sediments.

The congregation of the antibiotics extends from 1.01 μg/kg (sulfachloropyridazine) to 485 μg/kg (sulfamethoxazole). Under appropriate environmental conditions residue of antibiotics on sediment effects, microbial communities and only individuals who are resistant can survive.

3.9. RELEASEMENT OF SUBSTANCES WITH INHIBITORY EFFECT TO THE AQUATIC ENVIRONMENT

Persistent utilization of inhibitor substances in aquaculture and substitution of likely bioactive substances to the aquatic environment is terrifying. Just 20% to 30% of the antibiotic-treated feed is eaten by fish, while 70% to 80% of it passes straightforwardly in the aquatic environment.

Oxytetracycline decomposes quickly in the marine environment however a large portion of them attach to the small particles and amass on the sediments. Even after 12 weeks, amassed oxytetracycline could still exist at a concentration to exhibit activity. Sediment that contains antibiotics affect the marine fauna adversely.

For instance, oxytetracycline residue was identified in blue mussel (*Mytillus edulis*) 80 meters from the confines where the antibiotic was utilized. While the antibiotics on the surface of the sediment lose their impact easily, antibiotics like flumequine, oxytetracycline, sarafloxacin, and oxolinic acid, within the sediment can be efficacious for as long as 180 days.

Antibiotics on the sediments become inefficient so easily not as a result of the debasement, however because of the spread. Antibiotics like trimethoprim and sulfadiazine have moderately less half-life within the sediment which is roughly 90 days.

On the other hand, the half-life of the florfenicol in sediments was determined as 4.5 days. Different synthetic and natural chemicals, for example, derris root, dichlorvos, and malachite green utilized in huge quantities in mariculture all around the world.

We may provide an example to disclose the aggregate amount of chemical utilization; in 1989, Norwegian ranchers utilized 3,488 kg of dichlorvos to control the outbreak of sea lice (*Lepeophtheirus salmonis*). Dichlorvos and the majority of the utilized chemicals cause grave effects on the environment and they are required to be utilized wisely.

3.10. CONCLUSION

Farmed fish is inconceivably resourced systematically, particularly when contrasted with other animal proteins (chicken, beef, and pork). The feed conversion ratio, which is the estimation of how much feed it requires to produce the protein, is 1.1.

This implies that basically, to produce one pound of protein, requires one pound of feed. Chicken, beef, and pork's feed conversion ratios range between 2.2 and 10. Thus, seafood's protein retention, as well as energy retention, are surprisingly high.

In comparison to wild fish, farmed fish are closely monitored, so farmers have more control over factors. This can positively affect the fish and the environment. Farmed fish are usually free of environmental contaminants like heavy metals and mercury, as they only eat feed processed by humans. The toxic level of fish feed is regulated.

The farming of filter feeders, for example, shellfish, can improve the quality of water. These creatures eat excessive nutrients in the water, which, thus, restrain the development of effluent. Filter feeders are frequently amalgamated into the farming of different species, like finfish, to utilize fish waste and uneaten feed as food, balancing the farm's environmental impact. This framework is called integrated multi-trophic aquaculture (IMTA) or polyculture.

Sustainable fish feeds are on the rise. Fish oil and fishmeal utilized in feeds may come from cuttings from processing plants. Moreover, substituting plant proteins for fishmeal in feeds is additionally getting more common.

The most well-known negative environmental impacts that have been related to aquaculture include water quality, water eutrophication, destruction or alteration of natural habitats, transmission, and introduction of aquatic animal diseases (FAO, 2006a).

Fish farming adding to marine water eutrophication, uneaten food pellets and feces from fish farms modify the organic matter in the sediment, which can alter the utilization of oxygen and cause local eutrophication

(Mirto et al., 2009). In 2005 Zaccone et al. explored the fish farming effects on both sediment and pelagic environments of Mar Piccolo of Taranto (Italy), studying the dissemination of microbiological markers (coliforms and heterotrophic bacteria), in relation to chemical and physical parameters showing that the benthonic environment escalated the bio-deposition of the sediments, causing changes in the plenitude and the composition (clostridia ratio/ heterotrophic microscopic organisms) of microflora. New strategies were created for the recognition of environmental indicators (Caruso et al., 2002a, b, 2003).

In expansion to the eutrophication that causes gigantic harms for the local fauna and environment, the outbreaks and spread of disease are unfavorable results of the diversification and expansion of the aquaculture sector related to globalization (Mancuso, 2013a, b; Mancuso et al., 2007, 2013; Crisafi et al., 2011; Amagliani et al., 2009).

The implementation and development of national aquatic animal health management plans are imperative to eliminate, prevent, and control diseases in a timely manner and retort to buyers expanding worries for food safety, animal welfare and ecosystems integrity.

Aquaculture requires an empowering policy environment to grow in a sustainable manner and to be incorporated into the coastal zone. In addition, the associations between aquaculture and the larger system where it happens, specifically, the impact of the encompassing social and natural environment on aquaculture, must be taken into consideration (Subasinghe, 2009).

REFERENCES

1. (2010). *Ecosystem Concepts for Sustainable Bivalve Mariculture.* [online] Available at: https://www.researchgate.net/publication/316660652_Ecosystem_concepts_for_sustainable_bivalve_mariculture (accessed 10 September 2021).
2. Global Aquaculture Alliance, (2019). What Is the Environmental Impact of Aquaculture? GAA. [online] *Global Aquaculture Alliance.* Available at: https://www.aquaculturealliance.org/blog/what-is-the-environmental-impact-of-aquaculture/ (accessed 10 September 2021).
3. Kaiser, M., Laing, I., Utting, S., & Burnell, G., (1998). *Environmental Impacts of Bivalve Mariculture.* [online] Heriot-Watt Research Portal. Available at: https://researchportal.hw.ac.uk/en/publications/environmental-impacts-of-bivalve-mariculture (accessed 10 September 2021).
4. Mancuso, M., (2015). *Effects of Fish Farming on Marine Environment.* [online] Fisheriessciences.com. Available at: https://www.fisheriessciences.com/fisheries-aqua/effects-of-fish-farming-on-marine-environment.php?aid=6896 (accessed 10 September 2021).
5. Michael, A. R., (n.d.). *Environmental Effects of Shellfish Aquaculture in the Northeast.* [eBook] Maryland: Animal Science Building College Park. Available at: http://citeseerx.ist.psu.edu/viewdoc/download?doi=10.1.1.1064.9142&rep=rep1&type=pdf (accessed 10 September 2021).

Chapter 4

Mariculture, Economic, and Social Impacts

CONTENTS

This chapter sheds light on the mariculture and its economic as well as social impacts. It highlights the socio-economic aspects, the contribution of mariculture to livelihoods, its economic role, and benefits, mainly its contribution to country's GDP, and the fisheries governance. It also talks about its function in social structure by addressing the marketing and export.

It puts emphasis on the definition of the social risk in mariculture such as the identification of hazard, the economic influences of the provincial mariculture. It also illustrates the gross domestic product (GDP), employment, and labor income, along with explaining the socio-economic impact reviews and value of mariculture.

4.1. INTRODUCTION

The way of life of marine organic entities requires the stock of ecological assets which change as indicated by the way of life species and technique. As culture strategies are different and huge and developing number of species are refined, it is unimaginable to expect to make numerous wide speculations on natural impacts.

In wide terms, non-carnivorous species have moderately low natural needs whereas carnivorous species may set elevated requirements; the way of life of autotrophs, e.g., seaweeds, is not considered here. On account of non-carnivorous species, e.g., bivalves, connections emerge from extraction of supplements that are generally traded from the framework in the product, in spite of the fact that there might be a nearby leftover effect through neighborhood grouping of wastes.

For carnivorous species in open culture frameworks, nourishment comes from outer sources, e.g., from agribusiness and catch fisheries, and the neighborhood climate is utilized straightforwardly for the handling of byproducts. The clearest ecological assets utilized for mariculture incorporate the arrangement of supplements and oxygen and the processing of wastes.

In case a whole system approach is considered then the environmental costs of constructing infrastructure such as nets, cages, boats, roads (often through remote areas) and buildings, together with the potential ecological costs linked with the transmission of diseases and parasites, the effects of escaped culture organisms and conflicts with wild predators, e.g., piscivorous birds should be included.

Artisanal mariculture is less significant in the marine climate than in new water so the fundamental drivers of mariculture development are monetary as opposed to social, even though the social results might be significant.

Both the financial and ecological parts of mariculture are best perceived for the enclosure creation of salmon in high latitudes and for the marine and bitter water lake culture of shrimps in low latitudes. Both culture techniques require huge contributions of outside supplements and carbon and both are basically cash crops depending on worldwide business sectors.

The giant tiger shrimp Penaeus monodon and the Atlantic salmon had ex-farm values of 3.86 and 2.16 billion US$, respectively in the year 1998.

Quite a bit of world salmon creation happens in moderately distant regions customarily overwhelmed by coastal fisheries and small-scale agriculture, exploiting a generally flawless, nonindustrial climate. Additionally, shrimp culture has used coastal regions customarily dependent on high quality coastal fisheries.

In the two cases, what was once in the past an asset held in like manner by those misusing catch fisheries has been moved to private possession, even though this is all the more regularly expressly an issue on account of shrimp culture in mangrove fringes. The advantages to far off networks from the work caused might be extensive in spite of the fact that, where organizations are not privately claimed, benefits may likewise be exported.

It can be said that mariculture is a rapidly growing or expanding and dynamic economic sector. It is literally making a significant contribution to the growth of various economies in the world.

Monetary improvement is broadly viewed as expanding human prosperity. The size of the positive effect of mariculture shifts hugely. It is making a specific commitment in helpless nations, for example, Bangladesh, and recently agricultural nations like Vietnam. Mariculture is likewise making a significant commitment, freshwater mariculture specifically, to the accessibility of nutritious food to both cultivating families and more extensive society (Figure 4.1).

Figure 4.1. An illustration of bivalves.

Source: Image by Flickr.

In opposition to what may be generally anticipated, the significant commitments of mariculture toward the improvement of human prosperity and value are to be discovered not food security in pay for helpless makers even though these might be significant for families engaged with oceanic farming.

Maybe they are to be found in upstream and downstream work in auxiliary service arrangement, and where mariculture is firmly economically situated and creates little, low worth species, in the improved arrangement of excellent creature protein for minimal expense to purchasers.

Mariculture generates employment for the poor, economic activity from the sale of high-value species in national as well as international markets, and low-cost fish for domestic consumption on local markets.

Advantages produced through the stock, handling, and conveyance chain can be considerable and essentially more noteworthy than those straightforwardly connected with cultivating.

It is not unexpected to peruse features that decry mariculture's negative impacts on the climate but then hard to track down reports about its significance as a supplier of livelihoods around the world. Mariculture outstandingly influences individuals and social orders a long way past clear commitments to food security or any positive or negative natural effects.

Around the world, 18.7 million individuals at present work as fish ranchers and, likewise with fisheries, this figure increments by three-to fourfold if secondary and postharvest business is incorporated (FAO, 2016). The pay acquired by every one of these utilized people upholds up to four dependents (Smith et al., 2010).

Expanded preparing of ladies and more prominent interest in the labor force have followed. Work figures reflect patterns of expanding production information over the previous years too. Fish farmers presently address one third of all representatives engaged with fish creation.

In the year 2000, 12.6 million fish farmers composed just one quarter of that global total. Macroeconomic benefits derived from export earnings are also self-evident (Smith et al. 2010), but these impressive numbers do not tell the whole story.

In any event, when mariculture exercises do not return a similar financial advantage for each unit exertion as fishing, mariculture work requests vary in a general sense from fishing and occasional (independent work) in this manner making particular benefits (Irz et al., 2007). Mariculture occupations offer a sureness of area, which permits fish farmers to settle on decisions about family position and lodging that improve family dependability.

This gets numerous benefits over fishing in terms of education, wellbeing arrangement, and proper lodging (Fatunla, 1996). Moreover, routineness of working hours permits people to consolidate further education and other valuable arranged exercises into their everyday lives (Slater et al., 2013).

While fisheries may offer better returns at times of plenty, mariculture returns are for the most part more unsurprising in both time and worth. With this benefit, each farmer can settle on educated monetary arranging choices and savings.

Mariculture can eventually cause unwanted societal effects when it produces boom and bust cycles or otherwise collapses, for instance, because of the disease outbreaks, food safety recalls, or natural disasters.

Similarly, rapid commercial mariculture improvement can affect more conventional social orders by, for instance, prompting expanded degrees of debt. Unsettling influences in societal flexibility and decrease in social capital can be related with shifts toward high-capital mariculture.

Asset clashes can quickly emerge when customary clients feel that mariculture is infringing on their "patch" (Orchard et al., 2015). In any case, a large portion of these characterized aggravations to conventional social orders are common for quick growing enterprises, not mariculture alone. Most bits of knowledge into mariculture's societal impacts come from developing countries. In modern countries, mariculture is known to bring occupations and foundation, especially to secluded country regions.

Numerous mariculture enterprises in developed countries experience the ill effects of low accessibility of high-paying jobs joined with an absence of suitably prepared staff willing to work in modest situations for low wages.

Regardless, work maintenance in detached regions settles local area design and drives secondary industry and administrations (Neiland et al., 1991). Sadly, barely any sociological investigations have been led on mariculture in developed nations.

The research focus remains around monetary and cultural struggle around asset use, environmental concerns, and potential recreational/leisure clashes. A lot of this data has prompted progressively onerous and expensive administrative reaction (Abate et al., 2016; van Senten and Engle, 2017).

In developed countries, increasing importance is therefore being assigned to social license, or "the demands on and expectations for a business enterprise that emerge from neighborhoods, environmental groups, community members, and other elements of the surrounding civil society" (Gunningham et al., 2004).

Objective, stakeholder-led concentrates on the social and financial effects of mariculture (both positive and negative) could help build up a superior comprehension of and thus more prominent trust in mariculture exercises in both developed and developing nations (Leith et al., 2014).

Researchers go about as impetuses for stakeholder dialogue when such examinations are led and emphatically influence social permit and science strategy when they adequately impart the different advantages of utilization.

An improved research effort to address the financial and social effects of mariculture in precise and extensive manners is unmistakably required. There is convincing proof that asserts mariculture to be a worldwide monetary powerhouse that gives livelihoods and can be a driver of positive social improvement.

It carries occupations to separated and oppressed regions in industrial and developing countries and all around offers critical cultural advantages as far as admittance to food, foundation, schooling, and medical services. However, a lot of relevant inconstancy around mariculture in networks' remaining parts (Stevenson and Irz, 2009) and numerous significant inquiries are as yet unanswered. It is significant for mariculture analysts, expansion subject matter experts, the individuals who work in worldwide turn of events and strategy producers to comprehend these advantages and impart them to the more extensive logical and research community.

Recently, articles related to economic aspects of mariculture (Tokunaga et al., 2015; Dresdner and Estay 2016; Huang et al., 2016; Kumar et al., 2016; Hernandez et al., 2017) have been published in the Journal of the World Mariculture Society.

These have mainly centered around creation of financial matters; just one tended to social welfare assistance related with mariculture from the viewpoint of consequences for money disparity (Haque and Dey, 2016).

4.2. SOCIOECONOMIC ASPECTS

There is no uncertainty that mariculture can have tremendous monetary advantages which may run into billions of dollars. A considerable lot of these items are offered to send out business sectors despite the fact that, for a few, there are huge internal business sectors. Any place the market, the commitment of mariculture is clearly tremendous and structures a huge extent of public pay for some nations.

The financial advantages at the nearby level are, nonetheless, exceptionally factor and there are numerous cases where mariculture collides with an assortment of different stakeholders including inshore fisheries, the travel industry, and navigation.

Diverse culture strategies may also cause struggle, for instance, where synthetic compounds are utilized in fin-fish culture, appropriate separations should be kept up with filter taking care of shellfish culture to forestall cross-defilement.

There may be serious conflict when local users of a particular common resource, e.g., mangrove forests, are not consulted prior to the development of ponds that may have serious implications for their livelihood (Figure 4.2).

The coastal zone is viewed as the most important on earth and it is not astonishing coastal mariculture is at the front line of such conflicts as it is a generally new industry. Numerous nations are presently moving towards coastal zone the board as a fair technique for distributing assets to keep away from clashes.

Figure 4.2. An illustration of mangrove forests.

Source: Image by Flickr.

This includes educating all stakeholders with respect to the natural necessities of each utilization type and setting up a system of guideline and moderation to secure authentic rights.

A significant number of the biggest mariculture enterprises are presently possessed by global organizations and there is a chance of the benefits being sent out from the producing region while the natural risk is not. Most nations are currently moving towards guidelines which improve the supportability of mariculture and it is by such guideline, along with checking and authorization, that organizations can be obliged to acknowledge obligation regarding local impacts.

Moreover, especially for sent out items, shopper pressure is getting more sophisticated and there are as of now crusades against items from areas that have poor practice. Even though this is a somewhat blunt tool, which depends on the accessibility of clear, adjusted data to be really successful, it is probably going to turn into a major point in controlling the production of cultured organisms from the ocean.

4.3. CONTRIBUTION TO LIVELIHOODS

To represent the reliance on marine assets, an expected 50% of Mozambicans depend fundamentally on fish for protein consumption. In Tanzania, up to

70% of protein consumption may contain fish from freshwater or marine origin.

Population densities along the coast are high (>50% of 21 million people in Madagascar; >20% of 51 million people in Tanzania), and many of them are intimately linked to the sea for food security and jobs.

Although precise data are lacking, surveys suggest that 400,000 to 700,000 fishers engaged in marine fishing in the region between 2004 and 2013 (van der Elst et al., 2005; WIOFish, 2013), and considering that there is a dependency ratio of about 7: 1 (UNEP, 2001), this means that almost 5 million people are directly dependent on fishing for their livelihood.

Reliance on fisheries differs among nations, and is most noteworthy in Mozambique, Tanzania, and Madagascar, and lower in eastern South Africa, where the economy is more assorted.

4.3.1. Economic Benefits and Contribution to GDP

On a world scale, fish production generally contributes 0.5–2.5% of GDP (Bene and Heck, 2005). In any case, the commitment can be a lot higher in developing nations, where fisheries assume a more focal part in financial improvement, destitution reduction and food security (often > 5% of GDP in numerous west African nations; Bene and Heck, 2005).

Most valuation measure just the worth of production of fish, however on the off processing, trade, and administrations are added (these generally fall in other sectoral accounts), the general commitment of fisheries can be a lot higher (Figure 4.3) (WorldFish Center, 2011).

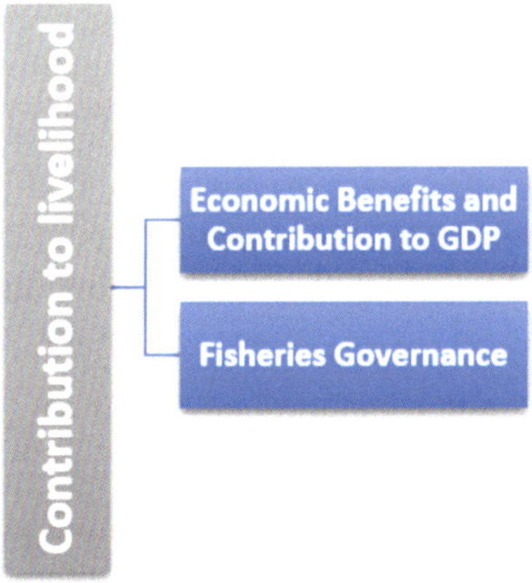

Figure 4.3. *Contribution to livelihood.*

Estimates of the contribution of capture fisheries to GDP (without any value adding) show exceptionally high contributions for small island states (30% for Seychelles and 15% for Comoros) and above the world average for Mozambique (4%) and Tanzania (2.7%). Contributions to GDP were much lower for Madagascar (1.4%), South Africa (1%) and Kenya (0.5%).

4.3.2. Fisheries Governance

Fisheries co-administration and the improvement of management establishments is a promising pattern in the SW Indian Ocean area. Beach management units (BMU) have as of late been set up in Kenya and Tanzania to co-oversee fisheries inside a framework with more extensive stakeholder cooperation (Oluoch et al., 2009, Japp, 2012).

Each BMU has authority over a geological region that establishes a fish-arrival point, which it manages mutually with fisheries office authorities. BMUs are engaged to impose charges against individuals for services provided, for everyday costs.

Even though a clear expansion to the social construction of fishing networks, it is not yet clear whether BMUs will prevail over the long haul. In Mozambique, co-administration happens through Community Fishing Councils (CCPs).

At the international level, SW Indian Ocean countries are signatories to many fisheries agreements, such as the 1982 UN Convention of the Law of the Sea; 1995 FAO Code of Conduct for Responsible Fisheries (CCRF); 2000 UN Millennium Declaration and Millennium Development Goals; 2002 Johannesburg Declaration on Sustainable Development and Plan of Implementation; 2005 Rome Declaration on illegal, unreported, and unregulated (IUU) fishing; and the FAO ecosystem approach to fisheries (EAF).

Tragically, most nations do not have the limit or foundation to successfully carry out the managerial and authorization tasks brought by these arrangements.

4.4. ROLE IN SOCIAL STRUCTURE

Social structure is characterized as an arrangement of geographically dispersed principles and practices that impact the activities and results of enormous quantities of social actors. The significance of catch fisheries in coastal networks as a wellspring of food and monetary movement makes it a significant determinant of social structure. Key actors are the fishers, mostly men in the proper fishing areas. Ladies, and in some cases youngsters, play a huge part in gathering shells, ocean cucumber and octopus in the intertidal for a couple of hours every day, normally during low spring tides, utilizing hands and sticks or rods (Jiddawi and Ohman, 2002).

They additionally use mosquito nets near the shore to get shrimps and little fishes. As well as gathering marine items, ladies assume a conspicuous part in the preparing and advertising of fish (Jiddawi and Ohman, 2002; Ochiewo, 2004). The 2012 artisanal fisheries outline review assesses that around 18% of the fishers are ladies. Different businesses that help fishing are artisans that make and fix boats and fishing gear.

Middlemen and traders play an important role in the artisanal fishery, by providing opportunities for fishermen who cannot afford to buy their own gear or vessels (Figure 4.4) (Jiddawi and Ohman, 2002; Ochiewo, 2010).

Figure 4.4. An illustration of trap fish ditch.

Source: Image by Pixabay.

Middlemen normally own gear (seine or gill nets) or vessels (dhows or boats with motors), which they lease to fishers. The cash acquired from the catch is ordinarily partitioned into three sections: one for the middleman, one for boat and gear support, and one for all fishermen on the boat, regardless to their number.

Dealers then, allocate the fish inland. Subsequently the actual fishermen get the littlest return. Fisher relocations along the East African coast is extremely old, and represents social transformation to a complex climate (Fulanda et al., 2009; WIOMSA, 2011).

Relocations are either brief or lasting, and pattern shift enormously, both inside and across country borders, and from a couple of days to a while or years. Drivers are a search for better gets and expanded pay. Fisher relocations are occasional, and related with social and monetary difficulties at home and host destinations.

4.4.1. Marketing and Export

The accomplishment of mariculture relies strongly upon the accessibility of business sectors for its items just as accessibility of data sources, for example, quality fish seed, feed, and water. Items are for the most part short-lived and require ice plants, cold storage, and appropriate vehicle facilities to guarantee their quality at the retail location.

These basics restrict farms to the area of urban centers. An exemption is seaweed, that can be sold and traded dry. The seaweed gather in Zanzibar is dried and offered to purchasers (brokers) who manage preparing organizations in the US, France, Denmark, and Spain (Msuya, 2009).

The purchasers have an imposing business model and set low costs for ocean growth. Farmers can store dry seaweed to build the cost by limiting supply, however towns rely unequivocally upon the pay. Value-adding in Zanzibar through handling has regularly been referenced as an alternative to build the pay from seaweed farming.

The principal fish marketing information services in the region are INFOPECHE (Intergovernmental Organization for Marketing Information and Cooperation Services for Fishery Products in Africa) and COMESA (Common Market for Eastern and Southern Africa).

These associations distribute a directory of fish importers and exporters yearly. Tough export prerequisites make small-scale producers less aggressive. To conquer potential advertising obstacles, producer associations can be framed to help farmers with technical and marketing viewpoints.

One such model is given by the local area-based culture of ocean cucumbers in Madagascar, where a set up organization of business and research partners give guaranteed admittance to markets, hatchery technology supply, and technical expertise.

4.4.1.1. Gaps in Capacity to Engage in Socioeconomics of Capture Fisheries and Mariculture

The clearest gap is an absence of hard data on the significance of catch fisheries and mariculture to the food security and neighborhood economies of families close to the coast. There is an absence of choices to catch fisheries, even, when catches from coastal regions are declining, and the populaces are expanding.

A typical pattern across a significant part of the SW Indian Ocean is to give better boats and gear to empower distinctive fishers to exploit assets further away from the shore. These approaches are, notwithstanding, being promoted without satisfactory information of deep-water fisheries.

A recent study (Everett et al., 2015) suggested that deep water crustacean stocks may be smaller than formerly thought in Kenya and Tanzania-it is important to investigate their fisheries potential before overcapitalizing on fishing fleets.

Mariculture, as a choice to catch fisheries, has not taken root as probable, apparently because of the constraints brought by the distance of the region, absence of framework and specialized ability. These obstructions must be defeated through a synchronized action by governments, private administrators, and NGOs. What this synchronized action should involve, and the boundary necessary to execute it, should be resolved first.

4.5. DEFINITION OF SOCIAL RISK IN MARICULTURE

This audit takes the point of view of the corporate area on social risk, i.e., "Social risks are challenges by stakeholders to companies' business practices over social consequences" (Kelly, 2005); and, with perceptions factored in, "Social risks are challenges by stakeholders to companies' business practices due to real or perceived business impacts on a broad range of issues related to human welfare-for example, working conditions, environmental quality, health, or economic opportunity.

The consequences may include brand and reputation damage, heightened regulatory pressure, legal action, consumer boycotts, and operational stoppages-jeopardizing short- and long-term shareholder value" (Bekefi, Jenkins, and Kytle, 2006). The rise of social risk is described by four parts in combination: an issue, a stakeholder or group of stakeholders, a negative insight about the organization and the way to do harm.

These basically equivalent definition of social risk made from a corporate perspective can be appropriately adjusted for mariculture at the sectoral, industry, organization, farmer group and individual farm levels as: Social risks in mariculture are challenges by society to the practices of the sector, industry, company, or farm over the perceived or real impacts of these practices on issues related to human welfare. The "polluter pays" principle exhibits this definition.

A farmer repays society through a tax or a permit charge for the expense of fixing harm from his contamination; or farmer assumes the expense by putting resources into a framework to avoid their activity from causing contamination. Something else, the farm could turn into the target of difficulties from the harmed community or from other concern group that see the harm and act on behalf of the community.

For example, the government authority could force a punishment or activist group could file a legal action. This definition also recommends three circles of social responsibility, which for the purpose of this audit are named internal, external, and global.

The internal circle would envelop duties to the farmer, farmer's family, and the homestead laborers (just as the cultured animals!); the external circle would be obligation to the local area in which it works, different clients of local area assets and the most general parts in the worth chain like providers, purchasers, and processors; and the global circle would incorporate responsibility to the remainder of the stakeholders, particularly buyers but also aquaculturists in different nations.

4.5.1. Hazard Identification

The wide and generally interlinked social and financial effects of risk incorporate loss of livelihood, loss of pay, loss of market, loss of resources and loss of ability to work productively. From this viewpoint, pretty much any danger can possibly convert into a risk that has social effect.

For instance, a natural disaster that not only wipes out the crop but also destroys farm assets and erodes the topsoil or silts up the pond will result in loss or severe and prolonged disruption of livelihood for the farmer and unemployment for the workers. Civil unrest, dangers to peace and request and far and wide neediness and social imbalances are by themselves social risks. Yet, these are not consequences of socially or environmentally careless acts of mariculture.

A farm or an organization choosing to situate in a space considered high-hazard because of social unrest is required to settle on a choice examination based on a definitely known hazard that could undermine the practicality of its tasks.

Additionally, farms or enterprise situated in a space where dangers of a social sort or origin are up and coming or unexpectedly happen would have to weigh management choices, for example pull out and keep away from the danger or stay and start risk management actions. This falls under project risk management. However, it is important project risk management incorporate a social danger evaluation, which could be a helpful strategy to embrace for investigation of dangers to mariculture.

It is informative in that an assessment of social risk to a project includes their effects for project expenses and reasonability. Besides, mariculture or some other financial area steers clear of bringing forth the most genuine risk of all, bad government, although opportunistic conduct from the business could abet it.

In any case, there are activities that farmers and industry can receive to improve the area's management and administration, including deliberate or self-administration measures and co-administration plans, producing unions with one another just as with different stakeholders like the science and innovation area, and sorting out into well-run professionalized farmers' affiliations.

Below is a list of social, economic, and political hazards to any economic activity:

- Civil unrest or civil strife;
- Social tension;
- Political instability;
- Rampant poverty (a proxy to weak government);
- High unemployment (an indicator of horizontal inequality between groups);
- Social exclusion (highly defined inequality in access to services and resources);
- Tendency of government to solve social conflicts by military action;
- Lack of independent judiciary (for dispute resolutions);
- Insufficient regulatory system;

- Excessive regulation;
- Poor or weak governance;
- Economic crisis.

The core of the definition of social risk that is, a challenge by the society to a practice or the practices of an entity-prevents such aforementioned conditions in risk analysis. This does not actually mean that they should be ignored, though their potential impacts can be very harsh and they are aided by improper practices in the sector. Small farmers who are much more vulnerable to such risks are required to be supported to deal effectively with them.

Another major category of hazards comprises of those that tend to prevent the farmers from adopting or to make risk-averse ones reluctant to adopt, parties such as an effluent treatment system or strategies such as crop diversification or intensification that efficiently enhances their management or livelihood.

Some of the examples are ill-defined property rights, seasonality or unreliability of labor, lack of protection of assets, perception of loss of profitability and a number of those listed above. Economic hazards that are laid in the market and industry, like changes in consumer preferences and tastes, appearance of substitutes, development of competitive products and market volatility, invariably translate to social risks.

The most severe result would be the collapse of a commodity industry and further the closure of farms. This would eventually result in widespread unemployment as well as the loss of income opportunities or livelihoods for communities and service sectors that are highly dependent on the commodity industry.

To be sure such group of hazards is not much prepared by practices within the mariculture sector, however the failure to recognize them could be attributed a range of reasons within the sector or industry, such as wrong interpretation of market trends, lack of foresight, or plain lack of capability for market intelligence.

In the case of technologies, the introduction of devices that displace workers in a social setting and where there is wide range of labor could reflect adversely on the farm's reputation. It could breed resentment from the community due to the loss of job opportunities (a similar challenge could be provoked by hiring practices). The same applies to the technology that needs higher skills which would replace lower or unskilled workers or labors. A farm or a corporate body that avoids training its workers and find it even more easy and convenient to replace them could produce the same response of resentment or direct hostile action from the community.

Worker relations, hirings, and purchasing practices stance a kind of social risk to the farm if these were observed by the community as exploitative, discriminatory, or opportunistic. In addition, there can be a case of offering competitive salary structures or wages to undercut competitors in the labor market.

This could eventually result in other sectors losing their workers or work force to the other sector or being forced to compete. The latter one would have a positive impact on the labor market of the community, but could also result in severe impacts on other sectors or industries and a fundamental feeling of ill will from the business sector towards the mariculture farm or company.

While on the other hand, a business strategy such as merger, consolidation, or acquisition that is meant to develop value for owners and stockholders-and could consequent in workers being made redundant-cannot be regarded as a hazard, notwithstanding this possible consequence.

The siting of farms, farm management practices like discharge and effluent treatment, and other mariculture related practices carry social as well as environmental impacts to the community. The environmental impact invariably translates to the social impact.

Conflicts can arise as access of people to the shore is blocked by mariculture installations, salination of crop lands, decline or encroachment in fish catch due to various mariculture impacts that include fish kills on the wild fishery (FAO, 2006).

A very basic example of a social hazard is the sitting and the practice of brackish water shrimp farms in India, which were cited by activists in their petition to the Supreme Court of India to shut down the sector in 1997 (Patil and Krishnan, 1998).

The use or application of inputs such as feed, chemicals, and drugs are a major source of social hazards, not much to the farm as to the industry or to the complete sector as a whole. A scare caused by a tainted product gives the industry invariably a bad press that potentially results in boycott or consumer resistance, importing countries' burning of containers of the product and also, change in the product or product-supplier preference, all of which led to the loss of market.

Loss of the market could threaten the viability of the sector and the welfare of workers and people dependent on it for a living. The return or burning of shipments of shrimp from Viet Nam found with unacceptable levels of residues of banned drugs also harshly impacted on the livelihoods of poor agricultural communities dependent on shrimp mariculture (MoF/NACA, 2005).

The exposure or ingestion of a farmer or farm workers to the toxic materials from drugs and chemicals because of lack of safety precautions can reach the media and become a serious local or national issue, having the potential of escalating into such challenges like lawsuits, community action against the farm or consumer resistance to the product.

The process of recognizing the hazards with social consequence includes posing an important or critical question "What challenges to the industry can be expected from society or certain stakeholders if something went wrong?" Answers to "What could go wrong?" which should be the first question, can be found or inferred from:

- Codes of conduct;
- Principles of good mariculture;
- Codes of practice;
- Good mariculture practices;
- International agreements;
- Certification schemes;
- Ethical and fair-trade standards;
- Animal welfare and free range;
- Labor standards;
- Rules and regulations;
- International Standards Organization (ISO) standards.

Such instruments can be efficiently used to recognize the hazards, which means to assess what could actually go wrong. Beyond this, mariculture sector needs to know what kind of issues or challenges can be expected from any sector of the society if anything goes wrong.

For instance, introduced species that become pests or the ones that carry pathogens have caused the collapse of fisheries and mariculture operations in some cases. This resulted in huge losses in revenue and severe implications for farmers, fishers, post-harvest industries and human health (APEC, 2003).

The risk analysis methodologies that are used for alien or introduced species are very well established and the methodology to assess or evaluate their economic, social, and environmental impacts have been developed. It is the likely challenges to mariculture as a whole or, for instance, the ornamental fish industry, if it were the source of the alien, that their impact would provoke that need to be recognized, evaluated, and mitigated.

The hazards that could actually provoke the challenges from different industries in various other nations are those having potential impacts from policies of the nation like subsidies or the target of the sector that is species and production targets, and marketing practices, for example dumping.

Subsidies, as well as protectionism, could cause severe impact to a similar industry and its workers in another country. Over-production as well as flooding the market therefore depressing prices would hurt competitors in poorer areas or countries, and dumping can create a lot of economic backlashes on an industry or commodity sector.

A study of shrimp farming in Latin America and the Caribbean by Wurmann, Madrid, and Brugger (2004) provides an instance with an interesting viewpoint. The study basically focused on two sources of competition: first one is producers in importing countries (such as the United States shrimp fishing industry) and second, producers in other regions, particularly Asia.

The study considered the anti-dumping case in the light of its negative impacts on national shrimp industries. It assumed that after the completion of the exercise that is, antidumping charges and imposition of countervailing tariffs, and countercharges, things would more or less go back to where they were at the beginning, but not before causing disruptions in producing countries, along with financial collapse of traders, distributers, and importers.

In addition, it also perceived the Asian competition on a large scale from the expansion of white shrimp production (*Litopenaeus vannamei*) as originally disruptive to the industry in Latin America but eventually concluded that it will induce the latter to become much more efficient in the long term.

Probably, the anti-dumping action did not affect the shrimp industries of the concerned nations as severely as the study assumed, although it created disruptions. While, on the other hand, there was no issue based on this from Latin America to Asian competitors (particularly those producing *L. vannamei*).

4.6. ECONOMIC IMPACTS OF THE PROVINCIAL MARICULTURE SECTOR

4.6.1. Gross Domestic Product (GDP)

Gross domestic product (GDP) impacts that are produced by the mariculture industry have largely increased over the past 10 years. In the year 2013, total GDP from the aquacultural activity, including direct, indirect, and induced activities, was estimated to be around $104.1 million. While in the year 2003, total impacts were estimated to be $10.5 million. Total GDP impacts from mariculture activity over the last decade were $476.3 million, with a compound annual growth rate of 25.7%.

The direct GDP benefit produced by the mariculture industry significantly increased between the year 2003 and 2013. In the year 2003, the direct impact was estimated to be $6.1 million. In 2013, that value increased to $66.2 million. The growth observed in direct GDP was mainly driven by annual increases in the volume of salmonid production, which grew by over 19,000 tons from 2003 to 2013.

In 2012, direct mariculture GDP decreased to approximately $26.9 million, as a major decline in the market prices for cultured salmon and large increases in feed costs reduced the return to capital. This decrease in direct GDP happened in spite of an increase in the volume of production from 17,264 tons in 2011 to 21,228 tons in 2012. Direct GDP in the year 2013 was increased by a rebound in salmon prices and high mussel prices, which eventually resulted in a significant increase in the total value.

Both indirect as well as induced GDP impacts from mariculture increased over the past 10 years. Indirect GDP, which is generated through Newfoundland and Labrador companies supplying goods and services to the mariculture industry, was around $26.2 million in the year 2013, up from $2 million in 2003. In addition, an estimated $11.7 million in induced GDP was created in 2013.

This compares to $2.4 million in induced GDP in 2003. Indirect and induced GDP growth shows expanded production volumes and increased nonwage operating expenditures for mariculture companies. Amongst the key support, service industries were equipment maintenance and repair, equipment manufacturing, packaging supplies, and transportation services (Figure 4.5).

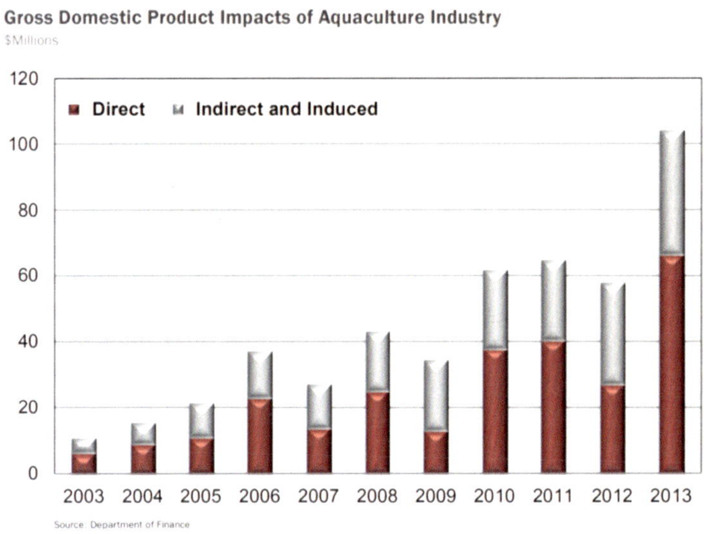

Figure 4.5. Gross domestic product impacts of aquaculture industry.

Source: Image by Department of finance.

4.6.2. Employment

Employment in the mariculture industry increased drastically between 2003 and 2013. Total employment impacts were estimated to be around 872 person years in 2013, as compared to approximately 225 person years in the year 2003.

Person year is defined as a variable that estimates the equivalent number of full-time positions that is produced by an industry. Though, this does not indicate the number of people employed in the industry. Employment impacts produced by the mariculture industry for the period covering 2003 to 2013 are shown in Figure 4.6.

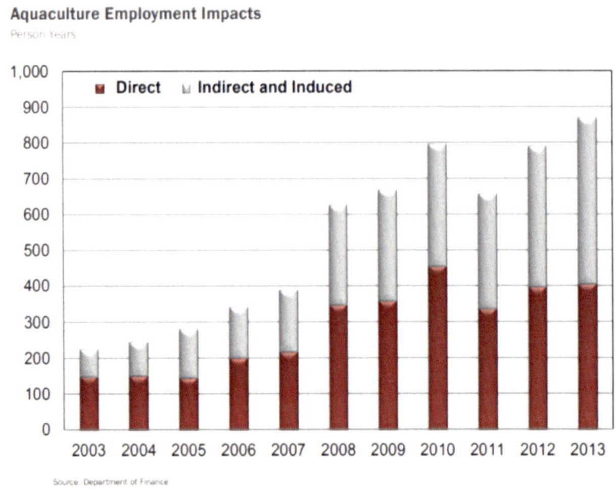

Figure 4.6. Aquaculture employment impacts.

Source: Image by Department of finance.

Direct employment in the industry of mariculture significantly increased over the past few years. Total direct impacts more than doubled, increasing from 150 person years to 406 person years between 2003 and 2013. Increased employment eventually resulted from year-over-year production volume increases. Spin-off industries in addition, also experienced employment gains between 2003 and 2013.

An estimated 465 person years of employment was created by spin-off activities in 2013, up from 74 person years in 2003. The growth observed in indirect as well as induced employment is also the key consequence of increased mariculture production that consequently resulted in higher levels of non-wage operating expenditures within the province.

4.6.3. Labor Income

Labor income impacts produced by the provincial mariculture industry have increased over the past few years. In the year 2013, labor income impacts were estimated to be $36.7 million. This shows a sixfold increase since 2003.

Total impacts on labor income from aquacultural activities were estimated to be $209.5 million over the period covering 2003 to 2013, with a compound annual growth rate of 20%. Labor income impacts for the industry are presented in Figure 4.7.

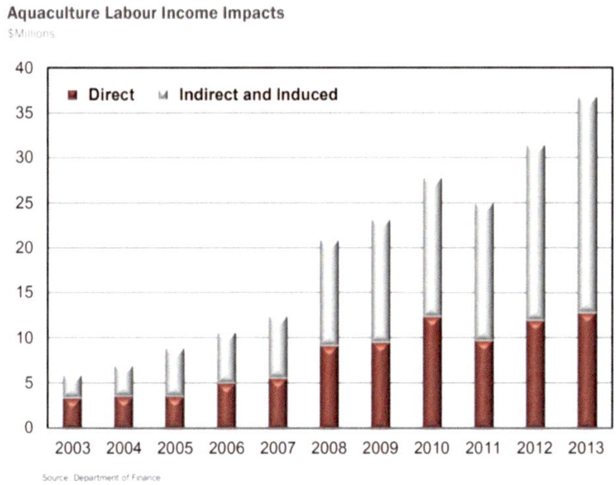

Figure 4.7. Aquaculture labor income impacts.

Source: Image by department of finance.

Since 2003, direct labor income impact has grown drastically. Estimates show that the direct labor income impact from the field of mariculture was $12.8 million in 2013, up from $3.4 million in 2003. This increase is mainly because of the higher production volumes, resulting in a higher demand for labor, as well as increased wages for individuals employed in the industry.

Average labor compensation per person year of direct mariculture employment was approximated to be around $31,400 in 2013. Spin-off impacts also increased significantly during this time period. In 2003, indirect as well as induced labor income was $2.5 million.

By 2013, spin-off labor income was estimated to be $23.9 million. Indirect income growth is mainly because of increased spending by mariculture companies within the province, especially higher levels of non-wage operational spending.

Induced labor income growth is largely the consequence of higher levels of consumer spending by those employed directly or indirectly by the mariculture industry. In the year 2013, average labor compensation per person year was around $49,700 for companies supplying the mariculture industry.

4.7. SOCIO-ECONOMIC IMPACTS OF MARICULTURE

A review of articles from a wide range of geographical areas highlights common social factors such as costs and benefits to the mariculture business and the wider community. Some of these benefits, include the use of mariculture to increase employment within peripheral towns or port areas where there are high levels of unemployment.

These days, mariculture is also being promoted as an activity that can effectively contribute to the conservation of specific habitats and thus maintain to certain level of biodiversity and is considered as a "legitimate and responsible partner with other natural resource users" (Burbridge et al., 2001).

4.7.1. Mariculture Industry Occupational and Functional Map

The structure of mariculture and mariculture farms is bit unique in that the majority are micro businesses and significantly employ people within the age group of 16–44 (Lantra, 2006). It is also the case that the majority of the proprietorships are held by men while the on-shore processing activities are dominated by women.

The sector of mariculture is regulated by a great number of directives that Lantra (2006) suggests could provide the potential for employees to obtain continued professional development. This would specially help in increasing their knowledge of regulations, legislative frameworks, environmental management techniques along with literacy and computer skills.

Neiland et al.'s (1991) review of the mariculture sector in Europe shows some social impacts to be considered. With reference to Scottish salmon farming, there is a concern that communities "switching away from more traditional occupations, may be gradually vulnerable to external financial strains and 'boom and bust economics.'

The benefits to the consumer were also emphasized because of the lower prices caused by increased supply of produce as well as improvement in the quality of the goods. In addition, with particular reference to West of Scotland, the sector of mariculture can significantly contribute to the stability and development of a community by offering work on a full- and part-time basis.

4.7.2. International Social Impact Reviews

In some of the case study areas, employment dependency of mariculture sector can be very high. As Burbridge (2001) notes mariculture in the do Ria de Arousa in Spain accounts for 25% of the local employment while in Finland there is employment dependency of 20.6% in Houtskari, 17.2% in Iniö and 15.5% in Föglö.

Where large farms can be developed particularly in the areas of rural communities and deprivation, employment is the key benefit. Although, over time employment numbers may not be effectively maintained or reach high levels because of the improvements in novel technologies that replaces labor.

According to Burbridge (2001), there are both positive as well as negative impacts of mariculture. For an example, due to the large initial capital investments that is needed for the setup of mariculture farms ownership is concentrated in the hands of a few; with the outcome that income distribution is skewed increasing income inequality and producing a subsequent negative effect.

In context to community stability, it is inferred from the studies reviewed by Burbridge (2001) that mariculture can provide several opportunities for parttime work that can further stabilize income amongst the workforce and hence, reflect the positive social impacts.

An additional benefit to the sector of mariculture is the restocking of some of the species that can be significantly used for the purpose of commercial tourism angling. Incorporation of mariculture with various other coastal activities extends the benefits of the sector to the broader community and also across other marine activities.

For so many years, mariculture has also been recognized to benefit not only farmers but also the wider community because of the investment in infrastructure and the demand for traditional skills for example boat building (White and Costelloe, 1999).

These elements within the local cconomy efficiently contribute to the decline in emigration as well as maintenance of traditions along with culture offering stability to these peripheral communities. White and Costelloe (1999) contended that these elements contribute to the maintenance of culture and identification of the rural coastal communities.

In addition, Carnerio (2011) reviews a number of papers on the social impact of mariculture highlighting in particular the papers of Bailey (1988);

and Primavera (1997) which themselves review the social impacts of intensive shrimp mariculture in the tropics and the Philippines respectively.

They note the following social impacts:

- Access to previously common resources is restricted due to privatization;
- Poorer people are further marginalized due to their inability to invest in mariculture;
- Labor-saving technology used in mariculture increases unemployment in the area;
- Rich has the ability to lobby politicians to mobilize resources to support mariculture;
- "Reduced food security, both from destruction of and limited access to coastal resources, and from disruption of local agricultural systems." (Carneiro, 2011, p. 355).

4.7.3. Value of Mariculture

There are several ways that can be easily followed to get the perfect insight into the kind of value that mariculture sector can provide to the community. These can widely range from the generation of gross value added along with the direct as well as indirect employment opportunities.

Comparatively the dependencies are quite small, for instance in South West England, mariculture supports 0.021% of the employment, followed by the North West which has only 0.010% of the work force employed by mariculture and Yorkshire and Humber at 0.09%.

Although, for the reasons given above, still this is critical to areas that are deprived or peripheral and restricted in the alternative economic sectors that could actually provide employment opportunities. According to the University of Strathclyde (2011), farmed shellfish in England decreased by a third in 2010 compared to 2009 figures totaling 3,283 tons.

In 2009 the total value was approximated to £7 million (CEFAS). Additionally, shellfish mariculture supports 135 full time equivalent jobs as well as 64 part time equivalent positions in England (CEFAS). In addition to that, ancillary businesses are supported by the mariculture sector providing further jobs and income into the local area "stemming emigration." (Burbridge, 2001). Additional studies are needed to extend the value attributed to the sector of mariculture. In specific, data on the gross value-added contribution of mariculture to NUTS region could provide additional support for mariculture.

Additionally, the value of significantly maintaining infrastructure connecting the towns with larger urban areas and spillover effect (like supporting the recreational anglers sector) should also be explored in a better way.

The enhancement in the skills of local employees and consumer surplus because of decrease in prices could also be valued to provide further detail of the direct and indirect and spatially distributed benefits to be gained from mariculture. According to Hossain et al. (2009), the non-monetary valuation of mariculture as compared to other marine uses has recently been used to select appropriate farm sites.

Hossain et al. (2009), used multicriteria evaluation (MCE) to identify the weightings given to particular factors of site selection such as water quality, soil quality, topography, infrastructure, and socio-economic by different stakeholders.

Such results would consequently enable better decisions to be made between alternative choice having conflicting objectives. No other study was found that elicited non-monetary values for mariculture in its own right within the UK highlighting the major potential for research efforts in this area, more specifically, to capture the socio-cultural impacts.

4.8. CONCLUSION

The social as well as economic impact of mariculture has not been subject to severe evaluation, and in general, the information is limited in this field. The available evidence shows that mariculture can efficiently generate significant social as well as economic benefits at a regional level.

The increasing competition and conflict between mariculture and other resource users requires a rigorous assessment of the possibilities for the integration of variety of sectors of the economy, and the evolution of government policy.

It is evidently the case that the important issues related with the mariculture development can he agreed upon by several interest groups, the major problem facing policy-makers is how to decide on priorities for action. Economists, together with other disciplines have a key role to play in providing relevant and reliable information on the interaction of important variables to assist decision-making policy evolution.

If the farm, an industry, or the sector as a whole obeys to socially responsible practices, it is fair to expect that it would face very little challenge or issues and none of them would be serious. Therefore, the need of the hour is to enable the farmers, traders, processors, input suppliers and others in the chain to adopt the codes, obey to better practices and comply with regulations.

A more specific challenge is how to avoid free-riding, corruption, and rent-seeking, and various other opportunistic behavior that would invite challenges to the sector with conviction. This shifts the focus of the issue to governance mechanisms, mainly the effectiveness of different mechanisms of governance (such as mandatory, market-based, and voluntary) instruments.

The other side of the problem is the capability of farmers to comply with a rising number as well as stringency of the requirements without jeopardizing their profitability; the main challenge is for farmers to see as sensible to business to adopt and comply with all these requirements. Various pilot studies and initiatives provide evidences of the effectiveness of such kind of strategies as organizing into farmer groups or more formal associations and adopting BMPs (Bueno et al., 2007a, b).

REFERENCES

1. Black, K., (2001). *Mariculture, Environmental, Economic, and Social Impacts of. Encyclopedia of Ocean Sciences*, [online] pp. 1578–1584. Available at: http://binet-repository.weebly.com/uploads/2/1/0/0/21005390/mariculture_environmental_economic_and_social_impacts.pdf (accessed 10 September 2021).
2. Jacob, O., (2016). *Social and Economic Impacts of Capture Fisheries and Mariculture* (pp. 297–308). [eBook] Inhaca Island, Mozambique: José Paula. Available at: https://wedocs.unep.org/bitstream/handle/20.500.11822/11349/rsocr_printedition.compressed_Part23.pdf?sequence=24&isAllowed=y (accessed 10 September 2021).
3. Matthew, J. S., (2017). *World Aquaculture Society | General Error*. [online] Was.org. Available at: https://www.was.org/articles/Societal-and-Economic-Impacts-of-Aquaculture.aspx#.YM5k6WgzbIU (accessed 10 September 2021).
4. MMO, (2013). *Social Impacts of Fisheries, Aquaculture, Recreation, Tourism, and Marine Protected Areas (MPAs) in Marine Plan Areas in England*. [eBook] Marine Management Organization. Available at: https://webarchive.nationalarchives.gov.uk/20140305104210/http://www.marinemanagement.org.uk/evidence/documents/1035.pdf (accessed 10 September 2021).
5. Neiland, A. E., Shaw, S. A., & Bailly, D., (1991). *The Social and Economic Impact of Aquaculture: A European Review*. [eBook] Belgium: European Aquaculture Society Special Publication. Available at: https://archimer.ifremer.fr/doc/1991/publication-3705.pdf (accessed 10 September 2021).
6. Pedro, B. B., (2008). *Social Risks in Aquaculture*. [eBook] FAO Fisheries and Aquaculture Technical Paper, pp. 209–228. Available at: http://www.fao.org/3/i0490e/i0490e01j.pdf (accessed 10 September 2021).

Chapter 5

Mariculture and Farming Technologies

CONTENTS

This chapter gives an insight to the mariculture and farming technologies. It begins with explaining the existing major mariculture species and faming technologies which deals with the marine finfish, seedstock production, and nursery.

It also discusses the grow out technology in detail along with illustrating the eight digital technologies disrupting aquaculture which further elucidates the cloud 3D printing save lives, robots will farm fishes or not, use of drones, use of virtual reality next generation of aquaculture.

As the chapter proceeds, it also highlights the integrated fish farming, backyard re-circulatory aquaculture system, along with explaining the farming in mariculture that is oyster farming, mussel farming, and clam farming.

5.1. INTRODUCTION

Mariculture, which is also known as aqua farming, was first started around 4,000 years ago in China. It started with the production of crap and now it is the fastest growing production in animal food has exceeded that of wild caught fish, and by 2030, aquaculture is expected to account for two-thirds of the fish that humans consume. It also comprises the production of shellfish, crustaceans, and seaweeds which actually gives the most important sources to humans, i.e., Nutrition, and molecular components.

The demand for fishes has grown at a very large rate which has put a shear on resources and sustainable practices among the fisheries, it basically requires the creative use of already existing and new technologies. There is a great way to produce the protein source as there is an advantage of technology.

Like other agriculture industries, the technologies have been introduced and there is an interest within the aquaculture as well in the farming industry. According to AgFunder aquaculture investment increased by 271% in 2016 over the previous 2 years.

The demand for fishes is evergreen because it contains a lot of health benefits for the consumers which is making them more interested in getting to know the advantages of their food. As a primary protein source, it is more efficient than any other sources by around six and four times. On a feed conversion basis there is a lot which can be done to improve the production and can bring efficiency in aquaculture.

5.2. EXISTING MAJOR MARICULTURE SPECIES AND FARMING TECHNOLOGIES

5.2.1. Marine Finfish

Marine finfish is established in the Asia-pacific region and is growing very rapidly. A very large range of species has been cultivated, and on the other hand the diversity of aquaculture is also increasing rapidly. In Asia the Japanese amberjack *Seriola quinqueradiata* makes up to 17% of the marine fin fish production, with only just under 160,000 tons produced in 2003.

Mostly all of the production comes from Japan, the production level there is relatively stable at 140,000–170,000 tons per annum since the 1980s. Other carangids that are becoming popular for this culture are the sub-nosed pompano *Trachinotus blochii*, and silver pomfret *Pampus argentus*.

Seabreams are a centerpiece of the Asian finfish mariculture production and a range of species are currently cultivated. Brackish water and mariculture environments are cultured in Barramundi or Asia.

Global production has been stable for around 10 years (20,000–26,000) tons per annum, though the production has declining in Asia and on the other hand it is increasing in Australia during this time. Grouper culture is increasing very rapidly in Asia but the prices are very high in the live fish market of Hong Kong and China and the wild-caught products are decreasing day by day because of overfishing.

Australia is using wild-caught juveniles and it has cultured southern bluefin tuna. Although the production of species is very low around (3,500–4,500 in 2001–2003) but the prices in Japanese market are very high as compared to others. It also supports a local industry in southern parts of Australia as well.

The production of 3,500 tons in 2003 was estimated around at $65 million. There is a species known as cobia which is engendering for the tropical region aquaculture. The maximum production recently comes from China which is totaled around 20,000 tons in 2003.

Moreover, the production of this culture is growing rapidly and now it is not only growing in Asia but also in some parts of America. Cobia is expanding at very increasing rate that it is now becoming a global commodity, in some or the other way salmon is also becoming the global commodity because of temperature aquaculture (Figure 5.1).

Figure 5.1. An illustration of milkfish.

Source: Image by Flickr.

Milkfish which is also known as chanos. It is having a long tradition of the mariculture there it is a very important part of the food. The major production of this seed is planted in Indonesia it is on a small-scale hatchery but are used for bait by Japanese in Indonesia.

In some pacific islands Kiribati, Nauru there is a culture of this production. Most of this production is brackish water ponds, this production is on an increasing mode in some large cages where the species are fed or the trash species are kept.

5.2.2. Seedstock Production

The hatcheries are producing a very large and wide range of the marine species but the production is still dependent on the fingerlings for the grown out which is somehow very difficult and costly to produce these such as grouper and napoleon wrasse (Sadovy, 2000; Estudillo and Duray, 2003) for this there is no technology yet.

The availability of these seeds is declining day by day just because of the overfishing and the destruction. There is a need for developing the technologies for production of the seeds mainly for the hatchery production.

In Asia there is a small-scale development of the hatchery production which only comprise of a few larval rearing tanks. The production of these hatcheries uses the very simple but the effective technologies so that the number of productions becomes large in number of seedstock a range of marine finfish species.

According to their tradition the most of the production is very wide that it includes more difficulty to rear these species such as groupers. On the other hand, the spectrum of the technology is very large for the hatchery production and it was developed in Australia and Japan.

Much of the hatchery technology in use in Australia has been adopted from Europe, and modified to meet local conditions (Battaglene and Kolkovski, 2005). The main thing to focus on these days is the technology in Australia for the hatchery technology (Figure 5.2).

Figure 5.2. An illustration of fish hatchery.

Source: Image by Wikimedia commons.

There is a need to reduce the labor for these technologies as it increases the cost. It has been declared that seedstock is the major production in China and it is as well as the center for the Asian pacific region, with almost 1,000 farms which are included in it.

The marine finfish production is highly specified in the production. For example, one farm may produce eggs from captive brood stock a second will rear the eggs, a third may rear the juveniles through a nursery phase (to 3–6 cm TL) and a fourth will grow the fish to market size.

5.2.3. Nursery

The seedstock is captured from the wild. Cannibalism is the main reason that some species are becoming endangered among hatchery-reared juveniles. There is a great loss of the transport because of these species.

5.2.4. Grow-Out Technology

Grow-out technology employed in the Asia-pacific regions ranges from small floating or fixed cages used by small family-run operations, to extremely large cages (15 × 15 × 15 m) used for amberjack grow-out in Japan or 30–50 m diameter circular cages used for southern bluefin tuna grow-out in Australia (Ottolenghi et al., 2004; Rimmer et al., 2004). Most of the marine finfish culture is from small to medium scale.

There are so many farms which only use very simple and easy techniques with some resources like wooden or bamboo cages and plastic barrels or polystyrene to provide buoyancy. Japan and Australia are those places which use the most sophisticated cage systems.

The technologies which are used in Australia are basically used in Europe. The Asian cage system is suited to sheltered inshore waters. The crowd in the coastal sites is increasing day by day a lot of countries are adopting cage designs which can mainly handle more water in it. These designs of the offshore cages are based from Japan and Europe (Figure 5.3).

More coastal area is opened up for farming because they have the option to farm in more open water. The main and major issue is the environmental impact on the continuation of proliferation of marine culture in Asian-pacific region.

Now there is a very strong understanding of the environmental impacts of the cage marine culture in temperature environments, a very little work has been done for these operations in tropical systems.

Figure 5.3. Fish nursery in Kandal Province, Cambodia.

Source: Image by Flickr.

5.3. DIGITAL TECHNOLOGIES DISRUPTING AQUACULTURE

5.3.1. Cloud 3D Printing Saves Lives?

Everyone can print their own hydroponic system and this is a truth! And this is very obvious that we can only print if we have a 3D printer. But there are a lot of people who still do not have a printer for their personal use. 3D printers are becoming more affordable, and it is possible that home 3D printers could become as ubiquitous as coffee makers in the future.

There is a company named 3Dponics which offers a lot of downloadable guides for printing in hydroponic system. This technology offers a lot of options like it can help in personal homes aquaculture gardens and can help in marine industry to produce hybrid aquaponic system.

Another example of 3D printing in aquaculture is a fish robot printed by MIT that almost shows an actual movements of the fish. This technology can also give a chance to further study and get to know more about the environment related to marine culture. This can result in better and clear understanding of fishes in their natural environment and it can improve the welfare conditions and provides a more experience in marine production as well.

As this technology gives a benefit that it also involves the production of the seaweed which gives it a very different sight. Aquaculture promises for causing disruption and even saving lives sometimes because it includes a lot of resources which comprise of water. Algae have been used to create an environmentally friendly, inexpensive material in the form of a gel used to 3D print medical implant devices (Figure 5.4).

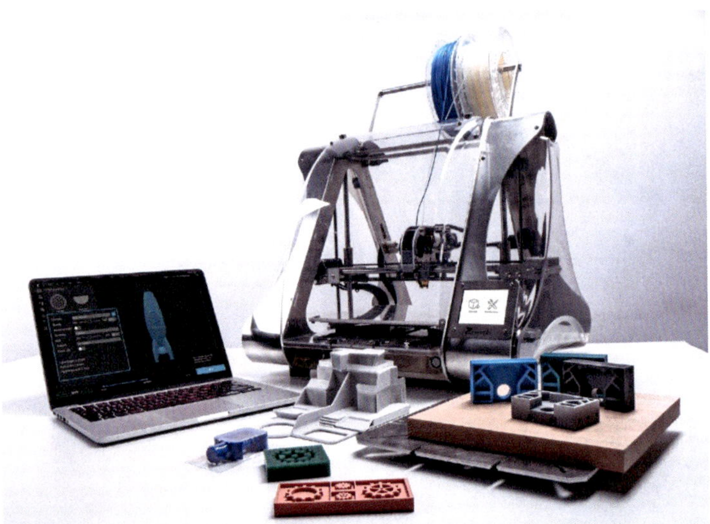

Figure 5.4. 3D printing technology.

Source: Image by Pixabay.

When the production increases the demand increases with it and it happens with companies as well for example Australian based Venus shell systems could reap the benefits. The main focus is on the human tissues and organs. Algae can also become a life-saving procedure.

5.3.2. Will Robots Farm Be Fish?

Considering a suitable alternative for wild fishing, the fishes which are farmed are not on their own. Fish farming is often in such conditions that it can create issues like some diseases and parasites, this can result in low yields and increase in the production cost. There is a well-known company which is working on this technology to help sick or the harmed fishes and for those also who are really willing to process in cermaq.

The future of fish farming could very well lie in giant, autonomous roaming robotic cages, called aqua pods, such as the sea station by innovasea. These cages have a very attractive look but they seem costly but it proves its efficiency against the stationary fish farms, protein from the sources of the fish has increased its demand.

What happens when an aqua pod grow fish in the open sea and the repairs are needed? There is a company named Norwegian SINTEF which is working on underwater robot which will be actually capable of repairing these nets, and will also provide a safer and the most effective way to handle this task.

How will we then get these offshore fish to market? This is believed by Rolls Royce that if robotic cargo ships are used then it will be the most clean, efficient, and cost effecting. This can also help in transportation of the fishes to the commercial entities.

In fact, rolls Royce has already signed contracts to transport construction materials for offshore aqua farms, though these will most likely be facilitated through usual cargo methods initially. There are a lot of other robots which are aiming to become a large solar power vacuum cleaner which can pick up around 150 tons of plastic from the ocean (Figure 5.5).

Figure 5.5. In coming times, robot will be used in fish farming.

Source: Image by Pixabay.

This technology can become a HUMAN AVTAR, which allows its system to work underwater and staying onshore. Maritime robots and deep trekker both provide the same function which is monitoring devices so that it can be used in exploration and aqua farming.

5.3.3. Drones Dare to Take on Dangerous Drives for Data

This is almost same as the robots; this also offers the same function as the robots which is it can be applied for aquaculture both above and below the water. It can be used in the offshore fish farming for example, and on the other side it can do any task which requires specialization and human resources, such as working underwater for repairing the damages and holes.

There are companies like Apium Swarm robotics which uses drones to survey the ocean. Using the sensor technology, it can have a complete survey of the ocean. There is another company which shows the live videos of the underwater exploration through an app in smartphone or with the internet. Companies like SeaDrone, Aquabotix, Powerray, and OpenROV are making very cheap drones for the underwater surveys for both professional and personal nature.

Drones provide a lot of options like it can collect information which can be used in a lot of surveys which can be later used in creating algorithms that develop the applications in the production of the aquaculture and the fish farms.

Sail drone, for example, offers the collection of the surveys and the fish stock analysis and the tracking of the environment and this can easily be applied for offshore aquaculture. This technology connects with a smartphone which allows for information that can be analyzed and gathered (Figure 5.6).

Figure 5.6. Drones dare to take on dangerous drives for data.

Source: Image by Wikimedia commons.

5.3.4. Sensors for Smarter, More Sustainable Aquaculture

The drones and the robots which are discussed above use sensor which can help to check underwater and collects data and the pH of the water, oxygen levels, turbity, and pollutants.

These are created by sense-T which usually helps in analyzing and checking a lot of things like the heart rate, the oxygen level, and water temperature. By this the metabolism can be measured as well. The farms of the shrimp use a sensorex to detect the levels of oxygen and the pH balance which creates a very ideal environment for the improved yield. Some technologies are very cool like we can say that are very simple and the easiest. One of it is the eFishery which is made to detect the hunger level of the species and after knowing this it becomes very easy to feed them on time. By this technology there can be reduction in the feed cost by around 21%.

Real tech company uses these robots and sensor to examine the quality of the water and it also uses the ultraviolet (UV) transmission to make the water pure and removes all impurities and hence there is a clean environment in the aquaculture.

Osmobet works on the land culture but it also allows for the management of the cloud and the connectivity of the smartphones. There is a special function in this which is it works on the hand sensing devices, feeding technology and transportation.

By using this there is an ideal atmosphere for the fishes. Other neat companies that offer entire monitoring systems which include Singapore, which offers real-time monitoring and connects for the cloud-based analytics, and Pentair, which offers a complete suite of sensor-enabled aquaponic device for small time hobbyist all the way up to commercial production companies.

The term known as artificial intelligence is basically for helping to make decisions in aquaculture. As collecting data is the main important source in the marine culture. A company in Australia provides a lot of technologies for each and every part of the Mainly the data is collected from the sensors which harness the actual power of artificial intelligence and helps in making the decisions fast and effective. A company in Australia provides a lot of technologies for each and every part of farming as it combines two functions together like sensing aqua which to create predictive analysis.

Shoal uses or swarm intelligence is a fish and robot (fish robot) it made up because it can help in checking the level of pollution underwater. These robots can make their own way in a group, together they can remove any obstacle which comes in their way it can be the other robot fish as well they can also recharge them as there are charging stations and these robots make decisions which humans could make.

There are a number of companies which are leading in market and use very basic technologies (camara and feeding systems) are now using artificial intelligence so that they can be competitive and can attract customers.

There is only 32% of the wild caught fishes which are cured unsustainably it is according to The Economist. There are a lot of benefits of using artificial intelligence like it reduce the number of overexploited species by camara and the collection of the data. It also helps the species to get better harvesting practices.

The latest innovation has launched a new platform which is made to help the researchers and scientists by which they can know a lot of new data and information. It is more effective in more control in the pathogens.

5.3.5. Augmented Reality (AR) Adds a New Dimension to Dives

The main use of this technology is that it helps in aquaculture industry. Already many countries like U.S. navy uses the DAVD which superimposes high resolution sonar imagery on a driver visual world. NASA has tested Microsoft's HoloLens in a similar way.

These contain scubas by IndieGoGo which is designed in such a way that it has a camara which allows calls between the divers. This technology helps in improving the efficiency of operations, and help in mortalities of the health status and an environmental parameter.

The best method to put AR in the aquaculture industry Is for the commercial purpose and somehow for the teaching as well. The Norwegian university of science and technology is designed as an aquaculture simulator using virtual reality and AR, incorporating oculus rift's technologies (Figure 5.7).

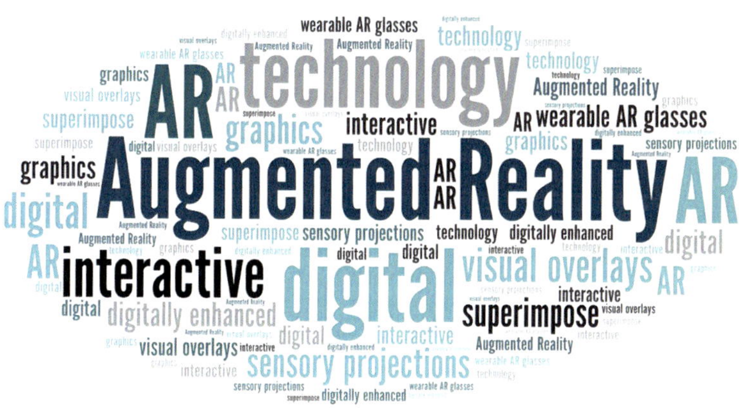

Figure 5.7. Augmented reality (AR) adds a new dimension to dives.

Source: Image by Flickr.

This program is made in such a way that it talks about the welfare of the fishes, prevention from the disease, and helps the fishes escaping from the dangerous conditions. This is known as the last concept and it is important for the students of Norway to be familiar with this concept of salmon farming.

5.3.6. Virtual Reality (VR) Is Opening the Eyes of the Next Generation to Aquaculture

This VR industry gives a lot of benefits mainly for training and education. VR principle is used for the next generation's interest in aquaculture it has developed that the uses of VR allow students to visit a fish farm online. This is said that these benefits can be used for the training purpose in marine culture industry.

5.3.7. Blockchain Verifies Sustainability, Improves Transparency from Fishery to Finished Plate

This is the best-known principle which basically secure the financial system like the methods of the payment which helps greatly in aquaculture industry.

The lifecycle of the fish involves the transactions, and the industry suffers from a poor reputation for the payments methods.

Blockchain is the online record which is incorruptible by one person. It is basically the opportunity for the transactions between suppliers and purchasers. To occur suppliers and safely. There is no need for the exchange of physical monies, and for adding the expense of transaction and the amount exchanges. More information is also provided about the harvest and the production.

These methods are used as well as stored and become more accessible to the customers of the production. The best and main benefit of this technology is the privacy it is important when these matters are discussed. Blockchain in made up in such way that the privacy is made and kept in mind like when claiming the fish.

Connecting all technologies together is the internet of things (IOT). The procedure of this technique is made in such a way that it revolves around the computing and the communications that makes it capable of doing the assigned tasks and the transfer of the information and data which is obtained by the sensors of the technology and it can analyze the smartphones as well.

For example, IoT technology, look no further that Eruvaka technologies or cargo zippers. The adaption and adaption of these eight digital technologies are non-stopping and are increasing day by day in many industries. This marine culture is usually a late adapter. But when we consider this industry, it becomes the fastest growing industry in the food production. The agriculture production of us has an additional 27 million tons of the fish production this all data is needed in the present because it is needed to maintain the level. Fish farming in the future will be more transparent and cleaner.

5.4. INTEGRATED FISH FARMING

This is divided into three stages; the first development stage is from the Han dynasty to the late 1950's; the second yield-pursuing stage is from the late 1950's to 1990's; the third and the most sustainable stage is from 2000 up to the future.

5.4.1. Yo Hu Bing

This is the earliest record of the fish farming which indicated that "yu," an official in the han dynasty build a fishpond south of Xian mount in terms of fan Li's first monograph of pisciculture in the world (473 BC). He was the one who planted this bamboo and the crops on very high dykes, in the neighborhood as well and he did water *Caltrop trapa* spp.

And Euryale ferox on the surroundings where there was water collected. It is a very different part of the comprehensive management of the farming of the fish and it is all done with the ecofriendly things like bamboo and the crop planting.

The actual shell of the turtle and the polyculture are already in the monograph of the pisciculture in the whole world. Monoculture are some common also like the crap Cyprinus which are already in some other positions because these are already replaced.

There is product which is known as the sheep pen and is built on the dykes, which also raise the sheep and sweep sheep droppings daily into the pond to the culture grass carp. The feces of grass crap in turn can feed silver crap. It can be transferred into an adult fish to big pond and dry the bottom to half dry and transparent grass *Nymphoides peltatum* as feed for new fish to grow.

This is a model of the fish grass which can rotate itself as well. In the 16 century the sericulture was developed in the rivers of the pearl. Some trees were replaced by the fruit trees like the mulberry and by replacing them with the dykes fish ponds were made. That is why these ponds are called mulberry tree dyke pond system.

5.4.2. The Development of Inland Aquaculture Depended Upon the Natural Seed Supply in Long Period of the Time

The collection of the seed and the transportation or it can be said that the pearl river after the tang dynasty. Though the supply of the seed was not enough. It was almost retarded and the development was made until the second turning point, it all happened in the late 1950s and the starting of the 1960s during the ongoing of the second stage.

On the other hand, China had a big achievement of the big head crap and the silver crap and also the white bream *Parabramins pekinenis*. This made the world record that a fish was able to spawned using hormones as it was first time in the history.

Then after this in 1960 and 1961 the black crap and grass crap were also able to spawns successfully though the breeding and hatching grass was tried in the river of the pearl in the very early 1921. But after this it solved many issues of the bottleneck problem of the seed as the supply was very less and it brought it to an end the collecting of fry from the great rivers. China also thrown the less supply of the seeds since the polyculture then it had a solid material foundation.

5.4.3. The Third Stage

In between the third stage the fish farming economically it was such as the crab macrophytes mollusks the system of the fish culture it was developed by the designs of the system. It was based on the five-kingdom theory principle. There were a lot of remedies for the eutrophic ponds water as it uses a lot of beneficial species like microorganisms, aquatic vascular plants, and some filtering animals.

The meaning of this is that the fish farming is very sophisticated and some systems are developing for this integrated fish farming system day by day. But the name of this integrated fish farming was defined in late 1970s by FAQ fisheries officials. As everything has a meaning fish also have a meaning of various aquatic products in China.

5.5. BACKYARD RE-CIRCULATORY AQUACULTURE SYSTEM

Recirculation aquaculture system is developed so that the water can be recycled and reused and after the water is filtered and all the suspended matters are removed this is called RAS technology. The technology of RAS is used for several purposes like for making use of water minimum in high-density culture of species. There is a very high approach than the other marine culture production system. As it is very obvious that new methods are better than the traditional ones the same is with this technology in past times growing fishes outdoors in the open ponds and raceways, in this system this technology is very typical for the fishes as they are reared in the indoor tanks and these are in a managed environment.

The water should be clean and filtered always that is why the process of recycling the water keeps on going. In this system both the mechanical and biological filters are used as this technology is based upon this, in this method a principle can be used which says that any species can live and can be grown in the aquaculture.

The main function is that new and fresh water is added to the tank to make it ready for splash out and evaporation just for removing the waste from the tank. The recycled water keeps on circulating in the system and 10% of total water volume is tanked is replaced daily.

In order to complete this procedure and to efficiently use in the capital investment in the recirculating system, the farmers of the fishes need to increase fishes as much as possible so that capacity increase. The farmers and the entrepreneurs also help the small scale to provide the fish facility in the urban and pre-urban where the resources are scarce, this all is done to promote the backyard recirculation marine culture.

These tanks are made to use the filters so that the water becomes pure and it reduces the waste and the ammonia so that the fishes are not harmed. When the water from the tanks is regularly recycled the water from the ponds and tanks are completely pumped out because it gets contain with ammonia.

The NFDB basically provides up to around 60% benefits and subsidy for the farmers and it helps them to set up new culture for the fishes. Their main aim is to basically increase the income from the agriculture. Farmers earn around 25,750 per month by just using NIRDPR.

The actual cost of a system is around 5.6 lakhs and the additional amount of around 1.4 lakhs is required for the electricity and other expenses. The NFDB provides almost 60% of the initial capital to the women as well as the farmers and 40% to the funding as well. This is said that some species of the fishes can be raised easily in this system.

This aquaculture system benefits a lot farmers who are living areas where the availability of water is low and there are some who are located very far away from the coastal regions.

This technology is basically used for the regions where the water is recycled and it is reused after filtration as it removes all the waste from the water. This is used for the high-density culture of species which are using minimum land area.

Technology partners are; national center for aquatic animal health (NCAAH), cochin university of science and technology, Kochi, Kerala

Ø **Things Required to Set Up as RAS Unit?**

- Land of around 100 sq.m land;
- Good water source;
- Source of feed and seed.

Ø **What to Fish Culture?**

- Most suitable for monosex tilapia, pangasius;
- Fingerling size (>2 gm).

Ø **What and How to Fish the Fish?**

- Pellet feed with 28–30% protein;
- 2–4 times a day;
- Manual broadcasting.

Ø **Objectives:**

- To encourage small-scale farmers and women to take up fish culture in the household backyard;
- To enhance fish production and consumption in daily diet;
- To promote income generation from small-scale fish farming and to improve livelihoods.

Ø **Who will Construct and Install the System?**

- NCAAH will assist the farmers in construction of fish tank and bring all materials and water testing kits, fabricate, and install the whole system including the cages, pumps aerators, filters, etc.

5.5.1. Resources

The most common and it is in dugout which are often large and are also fertile for the agriculture which helps them to convert into fish ponds for the main purpose. As the pond is large so it requires a large quantity of water which is needed to fill up the ponds in agriculture. The left-out water is untreated in the open. In the past times the fish farmers were only able to produce 2–10 hectare and used around 20 liters of water per kg of fish. An aquaculture needs only $1/8^{th}$ of a hectare and $1/6^{th}$ of water and by using only this much water and still able to produce 60 tons fish per year.

It is also highly efficient, it also utilizes the natural resources which is very limited which results in conservation of precious natural resources like water, land, and environment.

5.5.2. Driving the Youth Towards Farming

The system known as NIRDPR has discovered that modules are being established for its training program on the system of the aquaculture so that it can promote it in a sustainable mode of income.

"CUSAT has achieved the goal which was to operate a number of such ponds in Kerala." They are also planning to start installation in Telangana and Andhra Pradesh. There is one more benefit which is that it can be used in drought-prone areas where farmers can use supplements with a little income.

It is said by "Sakthivel that he mentioned about how youth is going to make return to agriculture a big challenge." This system is very easy to use and it has been maintained for past two-three months. The supervision of the technical field is required during the construction while this system is made and plans are afoot to train as well.

5.5.3. Precision Fish Farming

There are many objectives of this precision fish farming but the main ones are to improve accuracy, precision, and repeatability in the system. The second one is to facilitate more autonomous and continuous monitoring.

Third, to provide more reliable decision support. Fourth, to reduce dependencies on manual labor. This all means that PFF will improve the health of the animals and the welfare while it also increases the productivity. It also helps in sustainability in the commercial intensive aquaculture.

The definition of PFF clears that it helps in fish farming as cyclical operations processes realized and divided in four phases which are (observe phase) (interpret phase) (decide phase) (act phase) (Figure 5.8).

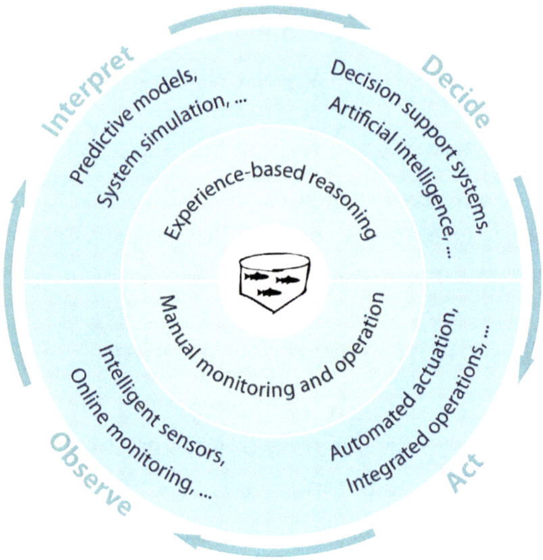

Figure 5.8. A cyclical representation of PFF.

In Figure 5.8, it is shown that maximum into different phases which are conducted by man. First, the farmer observes the fish via direct visual observation or with data acquisition tools such as cameras, the outcome of which is qualitative or quantitative information on the bio-responses of the fish. The farmer then uses some primarily this information, which is based on the current state and the conditions.

These are afterwards used as a foundation and helps to making decisions concerning farming operations and the managements also helps as the work is done manually on the cage. These decisions are based on the present situation and the expected future. It represents the manual functions of the feedback and feed forward principles in control.

Methods and tools for the farming of the fishes that apply technical solutions for the automation principles to one several or the different phases of farming operations may be considered PFF approaches. The result of using this system is that the operations will therefore for that the elements in that operation which basically which are shifted from an experience based to a knowledge-based regime.

5.5.4. Status of a PPF in Present Industry and Research

Though the concept of this has not changed and yet not designed, many technological research efforts and some researches are also made for the innovations and the fish aquaculture industry also comprise of some good tools and components for developing PFF methods, that are used in a few cases. There are some most relevant techniques which are used in single phase fish farming.

In this culture, the observations of animal are handled by the individual farmers. Though the innovations can automate the process, the foundation for the automation is already existing in the industrial foundation. It has established that it is for the variables for the animals.

It also has one more meaning that the development of new PFF in this area is considerable. As the fish farming has increased the production so now the research has extended for obtaining better results. The knowledge is also expanding of different sub-machines and bioprocess.

Before this knowledge is put on use or it gets start, it needs to be structured to give point to point information and data for the process which Is going on. There is mathematical model of dynamic system which basically offers and predicts that the system will respond on a given inputs and may also estimate features of the system that are said to be very difficult or impossible to measure.

In aquaculture result these mathematical models are made to measure the growth of a fish and its behavior. Such models are very good to PFF system as the help to predict the properties of the fish based on measured inputs.

These include the various types of the auxiliary data which is mainly required to drive the model dynamics but it also includes measured animal variables and this model can also convert it into feature variable which is even more useful.

For instance, a recent study is sought to estimate the economic yield of a production operation by combining a model of sea bream growth and with temperature as input with simulations of sales plans and strategies.

These models also represent the very important elements of the environment which helps in the production and its units also exist, these mainly cover some major subjects like the spatial and the temporal feed distribution which are gathered in the sea cages. If it is provided with sufficiently good input data, these can be directly estimated and are not associated with the fish but with the production.

The main use of the mathematical models is to just give an estimate of the states which are unabsorbed and are complex systems. It has a very long history within the control engineering and it is now seen the model is either based on statistical methods or just by using the non-linear methods.

These applications even allow the mixing of the two existing knowledge including the real time maker from the sensors which provide better estimates more than the possibility.

Since the fish production is a biological process. That is why it is more difficult to measure the different states and to process directly than in more technically oriented. This is the reason it is extensive and the use of the estimators are based on the mathematical models and it will be important that the procedure of the farming will require a better foundation for the information than it is available through the present virtual methods.

5.6. FARMING IN MARICULTURE

This culture is the most rearing of the aquatic species are in control and some are in semi-controlled conditions in the coastal and offshore waters where it is maximal and the subject is not related to daily and seasonal variations. Global production from the marine environment increased from 21.6 million tons in 1999 to 27.6 million tons in 2003. Mariculture contributed to 91% of the global total production from both marine and the brackish environment.

5.6.1. Oyster Farming

Oysters are one of the most valued seafoods and are farmed very largely. Nearly 12 species of the oysters are commercially famous and of these Crassostrea is the most important part of this farming. Around 4.2 million tons were produced during 2003 of which 3.6 million tons were from China and it also donated around 70% of this production.

5.6.2. Mussel Farming

It is one of the most important farming. And the most popular farming but only in the temperate countries. The production of the mussel has estimated around 1.58 million tons in 2003. There was an increase of 22.8 times over that of 1950, the contributions were made in around 40 nations which were producing nearly 9 species different to each other. Mussels have the anti-inflammatory, anti-histamine, and the research is still being going on for developing an effective low-cost mussel meat.

5.6.3. Clam Farming

Experiments are going on to form these species, indicated to the feasibility of this production in the pens or on the bottom methods. There is no culture of following calm strictly, but the method of the semicultured is to be performed by the farmers in certain areas to be harvested later.

5.7. CONCLUSION

This chapter sheds light on the mariculture and the farming technologies implemented in it. It starts with suggesting the explanation of existing major mariculture species and different farming technologies which inculcates and illustrates the marine finfish, seedstock production, and nursery.

As the chapter moves forward, it shows the importance and use of the grow-out technology. It then addresses the eight digital technologies disrupting aquaculture, which comes along with the describing the cloud 3D printing and its uses such as whether it is accountable for saving lives or not, or whether the robots are capable of farming fishes, the use of drones in deriving data, the significance of virtual reality, and how the blockchain is responsible for verifying sustainability, improving transparency from fishery to finished plate.

It then outlines the integrated fish farming technique along with highlighting the backyard re-circulatory aquaculture system. Towards the end, it illustrates the farming in mariculture which throws light on the different methods that is oyster farming, mussel farming, and clam farming.

REFERENCES

1. (n.d.). *Backyard Recirculation Aquaculture System*. [eBook] Available at: http://dof.gov.in/sites/default/files/2020-07/Backyard_Recirculation_Aquaculture_System.pdf (accessed 10 September 2021).
2. Alltech, (2017). *Eight Digital Technologies Disrupting Aquaculture*. [online] Eight digital technologies disrupting aquaculture | Alltech. Available at: https://www.alltech.com/blog/8-digital-technologies-disrupting-aquaculture (accessed 10 September 2021).
3. IndiaAgroNet, (2017). *Agriculture Technology | Backyard Recirculatory Aquaculture System | Ag Technology*. [online] Indiaagronet.com. Available at: https://www.indiaagronet.com/Agriculture-Technology/Backyard-Recirculatory-Aquaculture-System.html (accessed 10 September 2021).
4. Li, K., (n.d.). *Extending Integrated Fish Farming Technologies to Mariculture in China*. [eBook] Wuxi, China: Encyclopedia of life support systems (EOLSS). Available at: http://www.eolss.net/sample-chapters/c17/e6-58-08-10.pdf (accessed 10 September 2021).
5. Martin, F., (2019). *Precision Fish Farming: A New Framework to Improve Aquaculture*. Part 1: Global aquaculture advocate. [online] Global Aquaculture Alliance. Available at: https://www.aquaculturealliance.org/advocate/precision-fish-farming-a-new-framework-to-improve-aquaculture-part-1/ (accessed 10 September 2021).
6. Michael, A. R., (2006). *Regional Review on Existing Major Mariculture Species and Farming Technologies*. [eBook] Queensland, Australia: FAO: Department of Primary Industries and Fisheries. Available at: http://library.enaca.org/NACA-Publications/MaricultureWorkshop/Regional%20Thematic%20Review_Species%20and%20Technologies.pdf (accessed 10 September 2021).
7. Mohana, B., (2019). *New Fish Farming Tech with Smart Use of Water Aims to Double Agriculture Incomes*. [online] ThePrint. Available at: https://theprint.in/science/new-fish-farming-tech-with-smart-use-of-water-aims-to-double-agriculture-incomes/260692/ (accessed 10 September 2021).
8. TNAU, (2015). *Fisheries: Home*. [online] Agritech.tnau.ac.in. Available at: http://www.agritech.tnau.ac.in/fishery/fish_mariculture.html (accessed 10 September 2021).

Chapter 6

Methodologies Related to Mariculture

CONTENTS

This chapter revolves around the different methodologies related to mariculture. It sheds light on the identification and representation of the economic, social, and environmental costs as well as the benefits of aquaculture. It also gives insight to the reader about the culture methods in earth ponds.

This chapter provides a detailed overview of aquaculture methods and its practices. It also addresses the four common methods of aquaculture such as the pond systems, open net pens, and recirculating systems. It throws light on the recirculating aquaculture system by highlighting the advantages of recirculating aquaculture system, and constrains of recirculating aquaculture system.

Towards the end it talks about the basic principles of a recirculating aquaculture system along with outlining the intensive recirculating aquaculture system that is RAS.

6.1. INTRODUCTION

For raising the aquatic species there are many different methods on which the aquaculture can refer to. In the freshwater landlocked ponds, indoor hatcheries or culture tanks fish farming refers to the aquatic species, while in the salt water mariculture is aquaculture is occurring.

For the freshwater species and operations fish farming typically involves the aquaculture operations. In the fish farming catfish, trout, carp, and tilapia are common species grown, around the world which can be performed anywhere.

In the fish farming several methods are used, as it is considered as the most common task with the use of culture. To raise the fish from small fry to adult maturity large, plastic tanks can be used. To ensure the optimum water quality there is a need for the competent water recirculation systems, in terms of temperature and contamination.

To consume algae grown in the tanks the fish can either be fed high protein fish meal products or allowed to consume, depending on the species and system. In the fish farming manmade or the natural ponds are considered as common. Finally, in the raceways some aquatic species are grown that employ extensive water treatment and recirculation systems.

In the saltwater conditions mariculture occurs, where fish farming denotes to the freshwater operations. Mollusks and seaweed are the popular aquatic species in the mariculture. In addition to it, minor amounts compared to their fish farming counterparts with the crustaceans and finfish representing. For the Atlantic salmon the major United States is the mariculture industry, with the small amount of mollusks and shrimp. The China and the Japan are the considered as the major players, so it has been noted that Asia has extremely significant amounts of mollusk and shrimp mariculture production.

To the mariculture there are many methods. To raise fry or fingerlings in a hatchery is one of the oldest but probably the least efficient and then into their wild settings release them. On the ropes mollusks can be grown in the coastal areas.

For the fish and shrimp ponds are considered as the most common form of mariculture. It has been seen that from the nearby natural sources they are transferred to a pond that has been filled with saltwater, after raising the fish and shrimp to the fingerling developmental stage. In the saltwater estuaries and coastal areas off the coast of Nova Scotia, particularly the other fish salmon are raised in floating cages.

Every method of the aquaculture has its pros and cons. With the farmed fish center there are some issues and on controlling disease or damage to the fish themselves. In the ample concentration farmed fish live, regardless of the species than what they would naturally. For diseases or sea lice their close quarters allow to spread and it is considered easier in the wild conditions.

Moreover, between the fish their higher concentrations bring more actual physical contact, which will result in bruising and damage to their outer surfaces. It has been noted that than the farmed fish mariculture fish may be in less physical concentrations, and still, they are in higher numbers per acre than their wild cousins.

For fish this can result in higher concentration of their waste products, which even can have an impact on the habitat for the wild fish, only if in the case of mariculture occurs in pens in coastal saltwater areas. The higher risk of fish escaping their pens or ponds to live naturally is considered as another side effect of mariculture, also for the wild populations causing competition or genetic dilution complications (Figure 6.1).

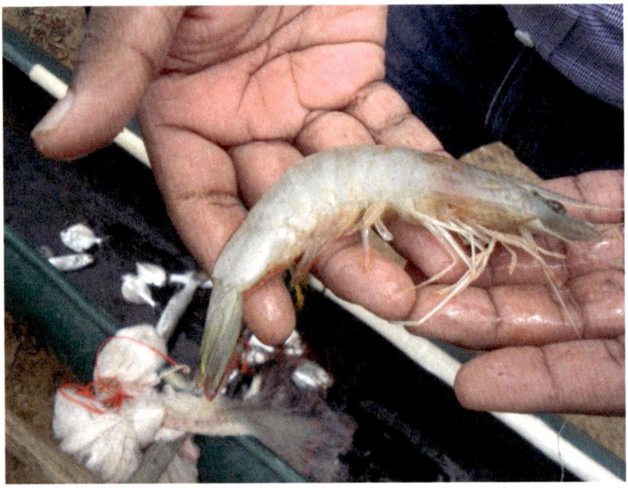

Figure 6.1. An illustration of white shrimp.

Source: Image by Wikimedia commons.

It has been said that significant economic, social, and environmental effects can be produce by the aquaculture production. The financial costs and benefits are being exceeded by these which are faced by the aquaculture producers.

For the development of integrated production models, a methodology is being proposed into the production management for the inclusion of the socio-economic and environmental effects of aquaculture. On a Social Cost-Benefit Analysis the methodology is being developed and it is being developed on the context which includes three parts.

First, with the environment that captures the interactions of aquaculture known as environmental. Second, for the incorporation of economic determinants in the production models that makes provision known as economic.

Thirdly, to the production and management process that introduces the social preferences is known as social. It has been noted that beyond pure financial metrics and beyond the quantification of private costs and benefits the methodology extends the assessment of the costs and benefits of aquaculture.

The development of the integrated model's aquaculture production is can also be supported and both the private and the social costs taken into the consideration along with the benefits associated with externalities and effects not appropriately captured by market mechanisms.

It has been seen that from an economic, financial, social, and environmental point of view the methodology can support aquaculture management and policies targeting sustainable and efficient aquaculture production and financing.

Rather than the profit maximization the aquaculture management often targets output maximization. This tactic is quite expensive but it can also be associated with social and ecological risks. Both the environmental and socio-economic impacts of aquaculture should be identified and monetized in order to achieve the goals of efficient and sustainable aquaculture development, with ultimate goal to be explicitly incorporated in the aquaculture production and management decisions. On the availability and the quality of the marine resources and the environment the aquaculture depends directly and also have the impact on it.

In this way it also interacts with the socio-economic parameters that the costs and the benefits of aquaculture extend beyond monetary expenditures and revenues. It has been seen that there is a general consensus that the sustainability of ecological and economic systems is tightly coupled.

Yet, it makes informed resource decision-making extremely difficult, and also the interaction among the latter is complex, at the time considering the dynamic nature of ecosystems and the difference in the scale of analysis of ecological and economic systems.

For the aquaculture the research on the development and the integration of ecological and socio-economic models is ongoing. There several conflicts which are there in these efforts such as the scale of analysis, the communication between ecology and economics, and the implicit assumptions employed, which have been identified in a way that explains the decoupling of these two disciplines.

> Recently, in the literature for aquaculture characterized by lower complexity as compared to the biological and ecological models alone several integrated ecological–economic models have been proposed. Into the bio-economic models there is some models integrating complex environmental and economic parts and linear models that can be categorized.

The benefits and costs of aquaculture is being systematically identifies, organizes, and valuates by the social costs benefit analysis (SCBA). At the end, in these stages the methodology is being developed. The cost and benefits are identified in the first stage. In the second stage they are being valued and quantified. It has been seen that they have a simple and transparent measure in a convenient unit, like market prices in monetary terms, social costs in contrast to the costs and benefits of goods and services and by the market prices the benefits are not always captured and in the material goods neither are limited to easily quantifiable changes.

Though, in a wider sense they should be regarded and quantified and along with this in individual utility and total social welfare measuring the changes.

There is proposed approach which aims at, benefits of aquaculture beyond pure financial metrics and monetary terms and extending the identification of the costs. And beyond the quantification of private costs and benefits and including in the analysis the social costs and benefits extending the modeling of aquaculture production.

So, in this way, both the private costs and benefits has been taken into the consideration, so that it can be produced integrated models of aquaculture production but with externalities and effects not appropriately captured by market-driven functions and factors also the social costs and benefits are being associated, so that producers internalize or can internalize, which can complement policies targeting aquaculture management and financing.

To the other production domains, the proposed methodology can be applied by moving beyond the aquaculture. Following adjustments have been altered to specific production-type characteristics. Among the other investment cost, production costs and revenues, social interactions, and environmental impacts the cos and benefits are being considered which are being translated into effects on ecosystem services and social effects.

As a result, to support the integrated management of aquaculture production a framework is being developed which allows for decisions that are consistent with the concepts of environmental sustainability, economic, and social efficiency.

6.2. IDENTIFICATION OF THE ECONOMIC, SOCIAL, AND ENVIRONMENTAL COSTS AND BENEFITS OF AQUACULTURE

On an aquaculture a rich body of the literature makes a lead that any attempt to assess its impacts should extend beyond production costs and financial returns. Extra efforts have been made to identify and quantify the costs and benefits of aquaculture which are being associated to answering questions related among other to the contribution of aquaculture to economic activity and income generation, for example, to gross domestic product (GDP) and employment the contribution of aquaculture and apart from providing food supplies the costs and benefits that aquaculture generates, such as rural development, regional stabilization and well-being, to the environmental effects of fish production, etc.

So, it becomes that evident which is the effects of aquaculture and the associated costs and benefits entail a plethora of economic, social, and environmental aspects.

In the terms of income and employment generation the economic effects of aquaculture can be identified and analyzed. In the recent years the contribution of aquaculture to world GDP remains limited, despite rising trends recorded.

Over the last decades in the world aquaculture has been the fastest growing food production sector. It has been seen that 50% of production came from fish products and 50% from crustaceans and mollusks, with regards to the EU, aquaculture production has been stable.

On the mussels the EU aquaculture focuses primarily (39% of total volume), trout (15% of total volume), salmon (14% of total volume) and oysters (8% of total volume). For more than 70% of total production the lead producers include Spain, France, Italy, Greece, and the UK, which together accounted for this. By France, Greece, and Spain (19%, 13% and 12% respectively), UK leads in terms of production value (21%), followed.

In the remarkable results the Mediterranean fish farming sector and its significant development have resulted not only with regards to the production of domestic fresh, cheap, and high-quality fish, but also with regards to the creation of a socio-economic structure that directly and indirectly involves a significant number of employees, particularly in the fisheries-dependent areas of the region (Figure 6.2).

Figure 6.2. An illustration of coastal area.

Source: Image by Pxhere.

It has been seen that from the other investment Mediterranean has developed to be the only productive activity that has colonized uninhabited islands and rock-islands which are normally excluded.

From the international experience studies on the costs and the benefits of aquaculture drawing show that employment dependency of aquaculture can be significant. For example, in the do Ria de Arousa in Spain it has been note that mariculture links to 25% of the local employment. In Houtskari, 17.2% in Iniö and 15.5% in Föglö.

Employment emerges as a primary benefit in Finland aquaculture employment dependency stands at 20.6%, especially in areas of deprivation and rural communities where large farms can be created. Though, due to technology improvements and subsequent labor replacement it has been found that over time employment numbers may decline.

With the large initial capital investments required additional economic costs and benefits of aquaculture are associated. In terms of the required investment in infrastructure and the demand for traditional skills aquaculture effects have also been identified.

It has been said that such elements may contribute to emigration reduction and maintenance of traditions and culture providing within the local communities, especially to the peripheral communities. In the maintenance of identity and culture there is an argument that these elements assist which is being made by the White and Costelloe.

It has been noted that from the public authority's aquaculture attracts much attention as a means of utilizing natural and national resources. At the regional level this driven by the significant social function of aquaculture, particularly in regions with depressed and marginal local economies. It has been said that not only for income generation but also for other social impacts the coastal towns and cities are often reliant on aquaculture activities.

As the positive and negative effects, the social impacts can be defined that aquaculture has in terms of: way of life, culture, health, and wellbeing. There is a report by the name Marine Management Organization, in which it is noted that social impacts vary by occurring scale (individuals, communities), time (current, future) and results (positive, negative).

From the increased food security, biodiversity, and way of life the communities and societies get benefit from. It has been seen that in better quality of life and wellbeing of coastal communities these benefits can be resulted.

From the experience of the UK the Marine management Organization drawing shows that the structure of aquaculture farms is quite unique as the majority are micro businesses employing mainly people at the age range of 16–44. Along with this same studies also resulted that majority of the proprietorships are held by men. Moreover, on-shore processing activities is being dominated by the female employment.

In the Europe Neiland review, the aquaculture sector which aim at identification of the associated social costs and benefits. There is the discussion regarding the concern that communities that switch away from traditional occupations may be vulnerable to external shocks with reference to the Scottish salmon farming.

As a result of lower process and the higher quality the benefits to the consumer were also being identified. In addition to it, through the provision of full-and part-time job opportunities they also note that aquaculture offers community stability and development with regards to the West of Scotland.

In other studies, this impact in the terms of community stability is also inferred, where it is recognized that aquaculture can provide opportunities for part-time work, and along this also provide the income stabilization and beneficial social impacts. For the commercial tourism there is an additional benefit of aquaculture which has been identified to be the restocking of certain species that can be useful. With the wellbeing and consumer preferences the others wellbeing have been associated and along with this consumption of goods that are produced at close proximity.

It has been seen that with specific environmentally friendly methods or food input, satisfies increasing consumer demand for eco-friendly produced food the availability of the fish varieties grown locally.

On the ecosystem impact studies to the have focused with regards to the environmental effects of the aquaculture, and along with this waste generation and their effects on water quality, marine life and, ultimately, consumer health.

As the date is being presented, so a schematic representation of the main environmental effects of aquaculture identified and discussed. Around the farm sites into the natural environment aquaculture can result in significant disposal of organic waste and contaminated water.

In the surrounding waters this can result in algae growth that can prove fatal for certain marine animals and indirectly constitute a danger to consumer health from consuming contaminated fish. In order to make room for intensive aquaculture there some other cases in which coastal ecosystems may face serious threats as they are destroyed.

To the impact of the farmed fish on local wild fisheries and endangered species there is another environmental outcome of aquaculture which is associated. In the farms disease and parasite outbreaks and can spread among farmed fish but also among wild fish populations.

Though, with the use of antibiotics and other chemicals in fish feed the disease outbreaks can be combated, moreover on the surrounding ecosystems concerns arise as such actions can pose serious threats but also on consumers' health.

On the aquaculture there are some environmental concerns which are further relate to the possibility of farmed fish escapes where in such case competition for food with wild fish can lead to their displacement. On the characteristics the selection of farmed fish is often based that make fish unsuited to living in the wild.

However, on the natural environment a certain number of fish may escape and put pressure and the wild spices impoverishing the genetic heritage and exacerbating the struggle for survival of native species. With the wild fish escaping farmed fish may interbreed and can affect the gene pool of the wild stock.

On the farmed species in many cases of aquaculture genetic engineering techniques are used, that are often not subject to external controls. At sterilization research on this sector has been growing rapidly aiming speeding up growth rates and increased resistance to cold and disease.

On the human health till the date on the effects of genetic engineering techniques is known. Therefore, in natural environments their impact on marine environment has been studied more thoroughly and research suggests that the sudden presence of this type of aquaculture represents a potential disaster.

It has been seen in many cases that to feed the farmed fish high quantities of forage fish are used. From a commercial point of view these practices make sense as the farmed fished command much higher prices than the fish used to feed them.

However, there are two closely related problems that emerge, the first one is the forage fish, such as sardines, mackerel, and herring, are also edible and second it might be the case that ultimately the quantity of fish used for feed exceeds the quantity of fish produced, putting high pressure on wild fish stocks.

To the water and energy use the additional environmental concerns are there and on climate change there is the impact of aquaculture. In the high demand both water and energy remain limited resources in aquaculture.

It has been seen that land-based ones require important amounts of water and electricity in the aquaculture system. To the CO_2 emissions of aquaculture processes and stages and their impact on climate change energy use is further associated.

In their calculations studies to the date vary in what processes they chose to include at the time of estimating emissions. For example, from energy use in feed blending, from energy use in transportation of feed materials and compound feed, from the manufacture of feed packaging it may be considered emissions.

On the environment some aquaculture operations have been identified having a positive impact, as well as on the human health. Indicatively, since they eat exclusively human-processed feed for which toxin levels are regulated, farmed fish are generally free of environmental contaminants such as mercury and heavy metals.

In addition to it, water quality can be improved by the filter feeders farming. As an activity aquaculture is being viewed which can contribute to the conservation of specific habitats contributing to biodiversity maintenance. It should be noted that the intensity of production and on the location of the farm the nature and magnitude of the environmental effects of aquaculture depend in large upon the species farmed.

6.3. CULTURE METHODS IN EARTH PONDS

In management and financial outlay there are three main methods, which also represents gradual development in complexity and of course coupled with higher production:

1. **Extensive:** It has been noted that extensive is considered as the easiest form of culture where no fertilizer or supplementary feed is added and the production from the pond depends entirely on the primary production which in turn depends on sunlight and nutrients available in the water and soil.

So, this is the reason behind keeping the stock density low and the recurrent costs are the lowest. In this method, in the general platform area of the pond it is essential that the water level be maintained between 35 and 40 cm to encourage algal growth on the pond bottom for the organisms to feed upon, in the case of mullets, milkfish, and other herbivores or penaeids.

It has been seen that to promote algal growth the light intensity that reaches the bottom is not sufficient in the case if water is deeper. It has been noted that in the case if water level is reduced below 35 or 30 cm then the temperature is likely to increase and may reach lethal limits.

This method is popular mainly due to the low financial outlay required for the operation especially in the rural areas of the developing countries. In the initial stage of mariculture development in developing countries this method is useful where it is still new and trained manpower is inadequate.

So, this the reason why the training phase is important and at the time of this phase improper operations can lead to loss of an entire harvest. Though, extensive system can better withstand these shocks than the other two systems because of the low financial outlay. In addition to it, with the basic system adequate experience is very helpful in understanding the more complicated systems and helps in mastering the complicated system faster (Figure 6.3).

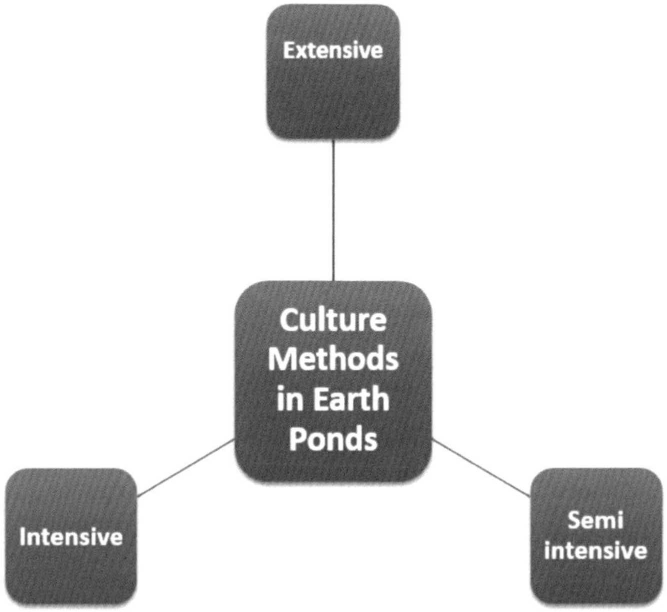

Figure 6.3. Culture methods in earth ponds.

2. **Semi-Intensive:** It has been said that from the pond the next step towards increased production is to stock at a slightly higher density and then to add fertilizers (organic or inorganic) and in turn increase the harvest. Within the pond the function of the fertilizer is to increase primary production which starts the chain reaction and ends in increased production from a unit area of the pond. In this method for the collection, storage, and application of fertilizer not only requires a slightly higher financial operational cost, but also involves additional labor.
3. **Intensive:** With or without the use of fertilizers supplementary feed is added to the pond in this method. In the pond the stocking rates would be much higher and also the depth of water greater than in the former cases because the food for the cultivated organisms is not dependent entirely on primary production within the pond.

Though, as a result the operating cost are much higher and the feed I quite expensive. With its greater risk the higher rewards also carry. A big financial outlay can be wiped out by an unforeseen calamity. At the initial stages where coastal aquaculture is still in its infancy and experience is inadequate, so for this reason and considering the heavy financial outlay required and along with this commencing with an extensive system would involve much lower risks and establish greater confidence.

6.4. OVERVIEW OF AQUACULTURE METHODS AND PRACTICES

In three types of environments worldwide a number of aquaculture practices are used for a great variety of culture organisms. In the fish ponds, fish pens, fish cages or, on a limited scale freshwater aquaculture is carried out either in rice paddies. In the coastal areas brackish water aquaculture is done mainly in fish ponds. For mollusks and seaweeds such as stakes, ropes, and rafts the marine culture employs either fish cages or substrates.

From extensive to intensive depending on the stocking density of the culture organisms the culture system ranges, the level of inputs, and the degree of management also. There some countries where the government focus on increase of production in the fish from aquaculture to help meet domestic demand, either as a result of the lack of access to large waterbodies (e.g., Nepal, Central African Republic) or the over-exploitation of marine or inland fisheries (e.g., Thailand, Zambia). It has been said for domestic consumption aquaculture practices are almost exclusively oriented toward production.

These practices include:

- Freshwater pond culture;
- Rice-fish culture or integrated fish farming;
- Brackish water finfish culture;
- Mariculture involving extensive culture and producing fish/shellfish (e.g., oysters, mussels, cockles) which are sold in rural and urban markets at relatively low prices (Table 6.1).

Table 6.1. Aquaculture Production Systems and Practices, by Region

Region	Major Culture Species	Major Culture Systems	Major Culture Practices	Scope for Future Development/ Needs for Further Expansion
Asia	At least 75 species; diverse freshwater and marine species, including high-value shrimps, mollusks, seaweeds, with carps and seaweeds dominating production	Traditional extensive to intensive	· Fish ponds · Fish pens and fish cages · Floating rafts, lines, and stakes for mollusks and seaweeds	Development of culture-based fisheries in inland lakes, rivers, floodplains, and permanent and temporary reservoirs and barrages
				Resource enhancement programs integrated with environmental management
Pacific	Mussels and oysters, red seaweeds	Intensive/semi-intensive to extensive	· Hanging lines for mussels and pearl oysters	Production of high-value species for select markets
			· Offshore cages for salmon	Small-scale aquaculture for local markets
			· Pond culture for shrimps, tilapia, catfish, milkfish	Improved management of fishery resources, particularly reef fisheries
			· Freshwater pens for crayfish	

Latin America	50 species of fish, crustaceans, and mollusks, including freshwater fish and marine shrimps in South America and mollusks in Central America	Extensive to semi-intensive and Intensive	· Offshore cage farming of Pacific and Atlantic salmon · Ocean ranching in Southern Ocean · Semi-intensive farming of marine shrimp in coastal ponds and extensive farming of freshwater fish in ponds	Production of species for export and marine shrimp and salmon
Africa	>26 freshwater fish; the most important being tilapia and common carp, mollusks, and oysters also	Mainly extensive, rural-based, integrated with poultry and animal husbandry, rice-fish farming; some intensive in raceways and floating cages	· Fish pond culture for freshwater fish · Raceways and floating cages for marine species	Increased emphasis on higher value catfishes for urban markets, on marine species of fish and crustaceans for select national market and export
				Culture-based fisheries in lakes and reservoirs
				Development of coastal lagoons which are almost totally unexploited
Mediterranean	>50 individual species, mostly freshwater and brackish water fishes-most important being salmonids and carps; oysters and mussels	Well-diversified modern practices, with highly technical and intensive systems in developing countries and semi-intensive and extensive elsewhere	· Fish pond · Fish cages · Ocean ranching	Production of high-value species of tourism and export Integrated coastal zone management

Caribbean	About 16 species of tilapias, carps, marine shrimp and, freshwater prawns, oysters, and seaweeds	–	· Floating cages in reservoirs	Priority is for aquaculture production for local markets
			· Fish pond farming in freshwater	–
			· Culture-based fisheries in reservoirs	–
			· Rope production of mollusks	–

Source: Table from ADCP Aquaculture Regional Profiles (1989b).

Low stocking densities are being used by the extensive systems and also there is no supplemental feeding, though to stimulate the growth and production of natural food in the water fertilization may be done.

Through the tidal means water change is affected which means only new water is let in only during high tide and the pond can be drained only at low tide. It has been seen that production is generally low at less than 1 t/ha/y. It has been noted that for the extensive culture the ponds are being used usually large (more than two ha) and may be shallow and not fully cleared of tree stumps.

As higher than extensive systems semi-intensive systems use densities (e.g., 50,000–100,000 shrimp PL/ha/crop) and use supplementary feeding. High densities of culture organism (e.g., 200,000–300,000 shrimp PL/ha/crop) and is totally dependent on artificial used by the intensive culture. For ease of the management both systems use small pond compartments of up to one ha in size.

By the applications of inputs semi-intensive and intensive culture systems are managed and by way of water management the manipulation of the environment primarily through the use of pumps and aerators. At the regular intervals the feeding of stock is being done during the day.

The computed daily feed ration is given in equal doses from as low as three to as high as six times a day in the intensive shrimp culture. On a daily basis water change is being affected with approximately 10–15% of the water in the pond replenished by the entry of new water in semi-intensive shrimp ponds.

Therefore, it has been said that semi-intensive and intensive culture systems are more labor-intensive than extensive systems which need little attention, and are costlier to set up and operate, not to mention the fact that they also carry higher risks of mortalities resulting from disease, poor management.

It has been noted that the production of course is much higher. From extensive culture financial returns are therefore much more attractive than those, though from semi-intensive culture studies have shown that the return on investment (ROI) is better than from intensive culture due to the high cost of inputs which is used in intensive culture.

6.5. COMMON METHODS FOR AQUACULTURE

It has been said that wild fish stocks are dwindling and there is an estimation which is made by the WWF that 53% of the world's fisheries are completely exploited, and the global fishing fleet is two-to-three times larger than the planet can handle.

By the middle of the century all the wild species could get collapse if one doesn't cut back on fishing soon. This means for meeting today's demand for seafood aquaculture is vital and from the fish farms one could find on the supermarket shelves.

In today's aquaculture landscape the following is a concise summary of four of the most common methods which are being used.

6.5.1. Pond Systems

Pond system is known as the oldest type of fish farming, which is originated thousands of years ago. With the clay-based soils pond farms include earthen pond, ditch or canal systems that can easily be diked to make enclosures.

On the today's industrial scale there are more sophisticated method which may be deployed but it is worth noting that pond systems are often depended on in developing countries, where a shortage of wild fish could have serious nutritional consequences. For several different types of fish, the fact that they are self-contained and can be used at once makes them an optimal choice for farmers.

At the present time ponds have wealth of potential when they are being handled properly. They are situated at inland, as from larger bodies of water and to be processed it is relatively simple to send the waste water into side pools, or used as fertilizer. The negative effects of pond enclosures are being significantly reduced on the natural environment (Figure 6.4).

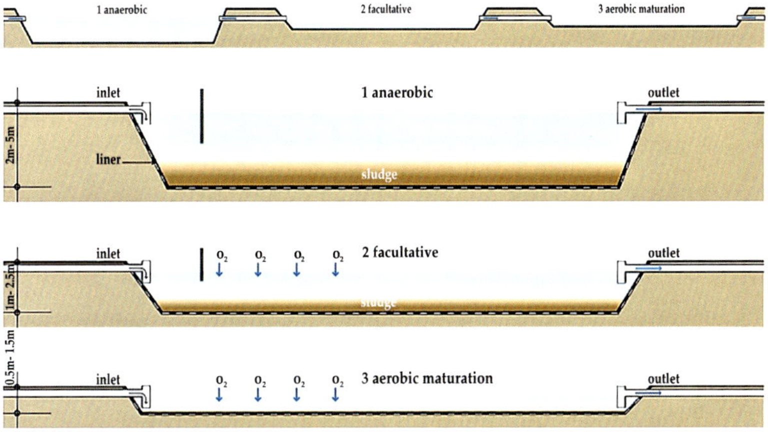

Figure 6.4. Schematic of waste stabilization pond.

Source: Image by Wikimedia commons.

6.5.2. Open Net Pens

In case of fish farming if one could have the read stories or seen pictures about it only then probably one has seen net pens. From the metal, wood, or bamboo a framework is being made which is involve in their most common form, which floats on the surface of a body of water and supports a mesh enclosure. To the bottom of the water body, they are either being anchored which is attached to a secure structure nearby, or both.

Because of their flexibility, versatility, and scale fish farmers favor the net pens. On whichever type of water body best suits the needs of their stocks they are being able to succeed. They are designed to allow water to flow freely through, which helps farmers deal with waste.

To reduce the pollutants reputable fish farms and processing plants have wastewater treatment facilities onsite. In an innovative ways wastewater can also be recycled, as it makes great nutrient-rich irrigation for agriculture, and can also be used to nourish other farm stocks such as shellfish, which are filter feeders, or catfish, which are bottom feeders.

To keep their stocks healthy companies that are certifiably sustainable do not depend on chemicals or antibiotics, which helps conserve local ecosystems. Additionally, within the enclosures floating fish feeds stay (Figure 6.5).

Figure 6.5. Open net pens.

Source: Image by Wikimedia commons.

6.5.3. Submersible Net Pens

These are said to be fully underwater, and often take the form of a large, closed-off cage instead of floating on the surface and an alternative to open net pens which are considered as the submersible net pens.

They are considered more remote than open pens since they are beneath the surface. From the elements this shelters them and makes it less likely that captive fish stocks will escape.

6.5.4. Recirculating Systems

For reducing the environmental risk factors there are some types of fish farming which show enormous promise. One example is recirculating systems, which are essentially like high-tech pond systems. It has been noted that in a series of interconnected fish stocks are reared, fully controlled tanks that are indoors and have virtually no chance of contact with wild stocks.

To the tanks a closed circuit of pipes carries clean water, while wastewater is pumped out of them, purified, and then recirculated. In the coming years, in the aquaculture industry recirculating systems might become more common.

6.6. RECIRCULATING AQUACULTURE SYSTEM

Water is being purified and reused continuously in a recirculating aquaculture system. It has been noted that recirculating aquaculture system is an almost completely closed circuit. There some of the produce waste products, solid waste, ammonium, and CO_2, are either removed or converted into non-toxic products by the system components.

With the oxygen the purified water is subsequently saturated and also returned to the fish tanks. To an absolute minimum the water and energy requirements are limited, by recirculating the culture water. So, it is not possible to design a fully closed recirculating system.

It has been noted that non-degradable waste products must be removed and evaporated water must be replaced. It has been said that till now recirculating systems are capable of reusing 90% or more of the culture water.

With the specific functions recirculating systems consist of a number of components, to ensure good water purification. So, the detailed explanation on RAS technology has been given below. From the system of water both systems are perfectly capable of removing the ammonia. On the many factors the choice of biological filtration is depending.

6.6.1. Advantages of Recirculating Aquaculture System

The advantages of farming in RAS are:

- Fully controlled environment for the fish;
- Low water use;
- Efficient energy use;
- Efficient land use;
- Optimal feeding strategy;
- Easy grading and harvesting of fish;
- Full disease control.

6.6.2. Constrains of Recirculating Aquaculture Systems (RAS)

There are a few constraints in respect of infrastructure, feeds, and staff, to allow the use of RAS grow outs.

To allow the use of RAS grow outs:

- Necessity for electricity 24/7;

- Good water source, preferably borehole;
- Good fish feed quality, preferably high protein, and fat extruded diets with high digestibility;
- Technically skilled staff able to work in a medium tech environment.

For the locations recirculating aquaculture system grow outs are the best option close to or in cities, with good availability of electricity. Next to this for farming tropical fish species in moderate to cold climates indoor by using RAS technology is the only possibility (Figure 6.6).

Figure 6.6. An illustration of recirculating aquaculture system.

Source: Image by Wikimedia commons.

6.6.3. Basic Principles of a Recirculating Aquaculture System

A basic recirculating system consists of the following components:

- Fish tank;
- Mechanical Filter;
- Biological Filter;
- Pump tank;
- Pump;
- Other items (e.g., UV-C light, oxygenation devices, aeration devices, feeders, monitoring, etc.) (Figure 6.7).

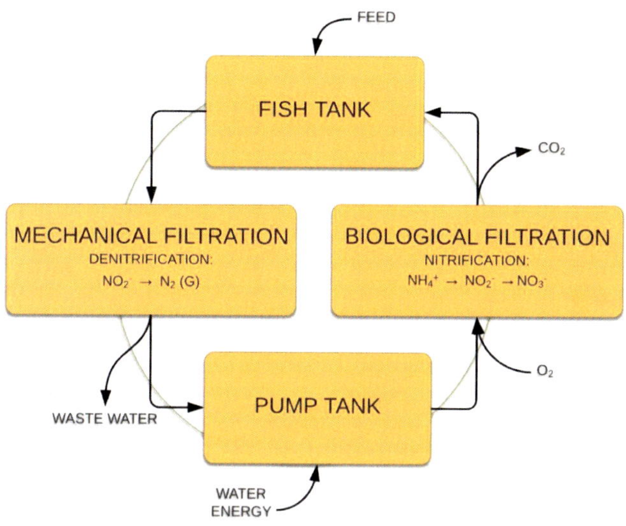

Figure 6.7. Basic principles of a recirculating aquaculture system.

6.6.3.1. Mechanical Filter

In a RAS system it has been seen that from the system water flow the mechanical filter is used to remove the suspended solids. To be removed in order to maintain water quality these solids, faucal matter is needed. There is choice for drum filter(s) or sedimentation filter(s) which is depending on the situation.

6.6.3.2. Sedimentation Filter

From the fish tank the water is flowing and it is led through the sedimentation tank. From the system water using gravity in this tank the non-soluble, solid particles are being separated. With a specially adapted polypropylene filter pack the sedimentation tank is filled.

On these filters the solids are being allowed. On the bottom of the sedimentation tank, it has the settled particles form a layer of silt. It has been said that this silt is biologically active and up to 60% of the nitrates produced by the biofilter are denitrified by the bacteria in the silt into nitrogen gas. On the feed system depending on the feed, the sedimentation tank must be cleaned regularly. In the farm the silt.

The water flowing from the fish tank(s) is in this tank the non-soluble, solid particles are separated from the system water using gravity. The solids are allowed to settle on these filters. The This silt is biologically very active: Depending on the feed load on the system, The silt in the farm. After this mechanical treatment, the system water flows to the next filter stage and water for cleaning is drained to the waste water sewage system.

6.6.3.3. Drum Filter

From the fish tank the system water is coming and can also be mechanically cleaned by using a drum filter. From the system water suspended solids are continuously and automatically removed and therefore no daily maintenance is required by the drum filter.

There is continuous need of the electricity by the downside of the drum filter. From the fish tanks the system water flow in the drum filter. It has been said that water is passed through the filter screen in the drum filter. It has been noted that by the filter screen the particles in the system water are blocked in this process.

The water level inside the drum will rise due to the clogging of the filter screen. As a result, into the operation the flush mechanism is put. Under high pressure the drum starts to rotate and from the outside of the drum water is sprayed from nozzles through the filter cloth, though it means flushing the waste particles from the screen.

In the gutter water particles is being discarded together with the waste particles. It has been depending on fish species and fish size the different filter screens can be chosen.

6.6.3.4. Biological Filter

By the fish NH_4 or the ammonium is being produced at the time digestion of the feed. For fish it has been noted that the ammonium is toxic. For the energy production ammonium (NH_4) is being used by the certain bacteria. In the bio tower of a recirculation system and transfer these bacteria are present into the nitrate and in an aerobe environment.

At the high-level ammonium is considered as toxic for fish, but it will cause problems at lower concentrations already. As a source of energy, the nitrate is also used for certain bacteria in the biological filter. In the relatively harmless nitrate these bacteria are being transferred.

There has been the creation of parts in a recirculating aquaculture system where these bacteria can grow in optimal conditions like the bio tower and moving bed filter.

6.6.3.5. Bio-Tower and/or Degassing Tower

To the bio tower the system water is being pumped first time from the pump tank which is also known as trickling tower or degassing tower. For the bio filtration a bio tower is consist of the polypropylene net filter blocks which is being suited for it.

Over the filter material on top of the bio tower the water is being dispersed. To the bio tower reception tank or directly back into the pump tank, the water flows through the filter packs. On the surface of the filter pack a biofilm of nitrifying bacteria is formed. To remove dissolved gasses from the water a bio tower can also be added to the system.

6.6.3.6. Moving Bed Filter

The moving bed filter is another biological filter. With the low volume these filters have a high filter capacity. With the aeration device the moving bed filter consist of a tank and it is also filled with filter beads. On the surface of the beads a biofilm of nitrifying bacteria grows.

6.6.3.7. Other Elements of a Recirculating Aquaculture System

There are many other items which can be added to a recirculating aquaculture system which is next to basic element which is being described above. There are some items like UV-C light, oxygenation devices, aeration devices, feeders, monitoring, etc., which are depending on the fish species items can be added to a recirculating aquaculture system.

6.6.3.8. UV-C Light

Below the visible light spectrum this light is electromagnetic radiation. It has been noted that wavelength of UV-C light is 200–280 nanometers.

UV-C light is highly germicidal and has an antiseptic effect. To a contaminated surface with the use of irradiation of UV-C light it is possible to control excessive growth of micro-organisms including bacteria. To kill the bacteria, viruses, and fungi UV disinfection is a proven method but first and foremost to keep the system water clear and free of small particles even at low dosage.

6.6.4. Intensive Recirculating Aquaculture Systems (RAS)

The temperatures and other variables can be controlled by the RAS offers the advantage and in order to maximize growth and maintain fish health. Some newer plants require only approx. 1–5% which is being compared to the flow through farms, minimizing the discharge to the environment and allowing operations to be established even where water supplies are quite small and all this is being required by the recirculating system.

36.6.4.1. *Generic Recirculation System Design*

All modern RAS systems generally have the following components;

- Pump;
- Bio-filter;
- Solid's collection systems (a rotating drum filter);
- A foam fractionation unit (also known as a protein skimmer or dissolved air floatation device);
- Carbon dioxide degasser;
- Nitrate filter;
- Sterilization points (usually UV sterilization);
- Tank drainage system that separates wastes from main outflows;
- Temperature control (heat pump or evaporative cooling);
- Heat exchange of effluent waters;
- Oxygen injection system;
- A pH control and alkalinity dosing system.

For the several components often, it is necessary to be combined for the simplicity and the capital savings. It has been said that resulting system is however generally compromised.

To build a RAS system there are many ways and no one way can be said to be best across all species, budgets, and situations. For particular application the choice of best system must be based on:

- Proven commercial production performance;
- Capital cost (feasible within the funds available);
- Operational cost;
- Reliability;
- Availability of backup and technical advice;
- Durability (useful lifetime of equipment);

- Ease of associated support activities (maintenance, feeding, grading, harvesting);
- Suitability for species cultured;
- Technology that fits technical ability;
- Redundancy/backup/unitization so plant can continue to operate if there is an issue with part of it;
- Flexibility to make changes as required without compromising production;
- Expertise and experience of the operator.

The one possible RAS design is being shown in the schematic below and to keep electrical costs low all pipe work is designed to minimize head-loss, e.g., pipes not limited in diameter and length and all bends are curved and not sharp elbows. From the sump to the tanks water is being pumped with either an aeration or oxygenation step on the way. To the drum filter water than flows by the gravity before returning to the sump. From the sump a second small pump draws the water and passes the water through a foam fractionator and returns the water via gravity to the sump.

In the modern high load RAS systems to increase the efficiency of carbon dioxide removal is being required in this system and a to the bio filter a second high flow volume loop is often installed which is parallel and work more efficiently with low to moderate flows but high flows are required to flush carbon dioxide from the rearing tanks and supply oxygen when stocking biomass per cubic meter is high.

6.7. CONCLUSION

Aquaculture was initially based on the raising of sea fish in fish ponds. Over the years, this method resulted in the copying and worsening of traditional farming problems to aquaculture, including:

- Outbreaks of disease and contamination;
- Massive energy consumption;
- Need for large areas of land and the consumption of large quantities of fresh water;
- Environmental damage;
- Ever rising growing costs.

These factors have seriously affected the quality of fish, the efficiency of raising them, and the narrowing of profit margins, sometime to the point that it is not worthwhile as a business. In some cases, the severe environmental damage has also delayed or prevented the use of fish ponds.

As a result, a new trend has emerged to raise fish close to shore. Most of this activity currently takes place in fixed cages at ports, bays, and fiords close to shore, protected from stormy seas.

Over time, the raising of fish in cages close to the shore or in bays has caused severe ecological problems in the bays and fiords. They include contamination, disease, increased fish mortality, and severe damage to marine ecology, quality of the environment, and tourism.

REFERENCES

1. Aquaculture ID, (2021). *Recirculating Aquaculture System or RAS-Aquaculture ID*. [online] Aquaculture ID. Available at: <https://www.aquacultureid.com/recirculating-aquaculture-system/> [Accessed 20 June 2021].
2. FAO, (n.d.). *Aquaculture Methods and Practices: A Selected Review*. [online] Fao.org. Available at: http://www.fao.org/3/t8598e/t8598e05.htm (accessed 10 September 2021).
3. JobMonkey Inc., (2021). *Aquaculture and Mariculture Methods-Fish Farming Methods*. [online] JobMonkey. Available at: https://www.jobmonkey.com/aquaculturejobs/aquaculture-methods/ (accessed 10 September 2021).
4. Land-Based Aquaculture Assessment Framework, (2021). *Intensive Recirculating Aquaculture Systems (RAS): Land-Based Aquaculture Assessment Framework*. [online] Lbaaf.co.nz. Available at: http://www.lbaaf.co.nz/land-based-aquaculture/intensive-recirculating-aquaculture-systems-ras-/ (accessed 10 September 2021).
5. Mariculture Systems, (2015). *Aquaculture Current Methods*. [online] Mariculture-systems.com. Available at: https://www.mariculture-systems.com/aquaculture-current-methods (accessed 10 September 2021).
6. Dillon R, (2021). *Four Common Methods for Aquaculture: The Healthy Fish*. [online] The Healthy Fish. Available at: https://thehealthyfish.com/4-common-methods-aquaculture/ (accessed 10 September 2021).
7. Sivalingam, S., (1981). *Mariculture as a Method of Using Coastal Zones*. [online] Fao.org. Available at: http://www.fao.org/3/AC583E/AC583E00.htm#:~:text=Mariculture%20is%20an%20activity%20involving,freshwater%20in%20the%20brackishwater%20areas. (accessed 10 September 2021).
8. Tsani, S., & Koundouri, P., (2018). A methodological note for the development of integrated aquaculture production models. *Frontiers in Marine Science*, [online] 4. Available at: https://www.frontiersin.org/articles/10.3389/fmars.2017.00406/full (accessed 10 September 2021).

Chapter 7

Mariculture-Sustainability and Regulations

CONTENTS

In the chapter, mariculture-sustainability and regulations, the importance of mariculture with respect to sustainability is discussed. The chapter also explains several mariculture guidelines related to biodiversity in which principles, standard, and certification is discussed. In addition, it also explains aquaculture regulatory framework. At the end, it highlighted development of sustainable methods. Some of them are lagoons/artificial wetlands, sustainable feed management, and sustainable use of chemicals.

7.1. INTRODUCTION

It is generally seen that assessing mariculture sustainability at the international level is an extremely challenging tasks, as information is not obtainable for every country with respect to the species that are grown, where that takes place, the methods that can be used, and what local environmental impacts and greatest sustainable yields could be expected for each species/method location combination.

The mariculture sustainability index (MSI) (Trujillo, 2008) was developed as an agenda for assessing the sustainability of aquaculture production using a score between 1 (poor) and 10 (very good). MSI generally includes 13 indicators, embracing ecological, social as well as economic aspects of the industry, for 359 country-species combinations, connecting 60 countries and 86 species.

The MSI score is attained as a combination of all 13 indicators. The Ocean Health Index used one subset of MSI indicators to compute mariculture sustainability and another subset to compute mariculture regulations. In addition, the indicator on genetic escapees was taken into usage to interpret the pressure of mariculture upon the conveyance of the biodiversity objective.

It is worth noticing that mariculture is becoming an encouraging area of aquaculture across the world and is one of the most imperative and rapidly fastest components of Asian aquaculture contributing considerably to the augmented demand for high value seafood items in the global market. India is considered as one of the largest leaders in the sector of aquaculture, contributing to about 5.2% of the total production in 2003 (FAO, 2005).

Globally, coastal aquaculture is one of the greatest developing food sector industries. The production from coastal ecosystem through farming that was less than 0.5 million t. in 1950, augmented to 10 million t. in 1990 and to 36 million t. by 2007 (FAO, 2009). A sub-continent, with seas all around on three sides, India has a long coastline of about 8,129 km. The country's mainland shelf is projected as 0.5 million square km, within its exclusive economic zone (EEZ) that ranges to 2.2 million square kilometers. The southern edge of the Indian peninsula extends in to the Indian Ocean, with the Arabian Sea in the west and the Bay of Bengal in its eastern part.

During the last six and half decades, the prospect of aquaculture for food production were extensively recognized and legal policies conscripted in several countries. The development and management of aquaculture is expected to decrease within the scope of several pieces of legislation and the capability of several institutions.

Aquaculture activities are required to be carefully controlled and monitored due to the fact that several interests involved, the diversity of natural resources used the number of institutions concerned, involvement of a wider range of stakeholders from both private as well as public sectors. FAO insist that States should create, maintain as well as develop an adequate legal and administrative framework that assists in the development of responsible aquaculture.

7.2. MARICULTURE GUIDELINES RELATED TO BIODIVERSITY (FIGURE 7.1)

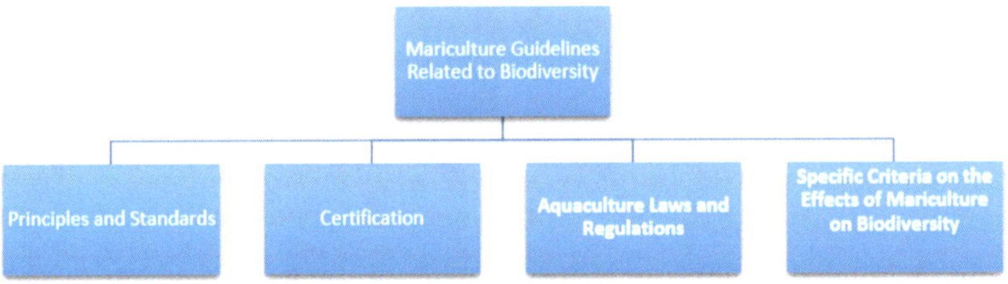

Figure 7.1. Mariculture guidelines related to biodiversity.

7.2.1. Principles and Standards

Although, no set of internationally approved standards has yet been developed particularly for the environmental regulation of aquaculture operations, many regional as well as national regulations and laws, based largely on scientifically accepted environmental criteria, have been accepted (Figure 7.2).

Figure 7.2. Equitable mariculture development.

Source: Image by Stockholm Resilience Centre.

Although, the ICES newly ready draft guidelines for the groundwork of environmental impact assessment (EIA) documents with respect to shellfish mariculture and the European Union funded the MARAQUA project that also offered scientific principles underlying the supervising of the environmental impacts of aquaculture. A number of standards and principles are voluntarily being applied to the industry in an effort to lessen its environmental impact and enhance its public image.

Moreover, in its decision VI/7 A, the Conference of the Parties to the Convention on Biological Diversity accepted guidelines for integrating biodiversity-related issues into EIAs. Article 9 of the FAO Code of Conduct for Responsible Fisheries (CCRF) offers a set of voluntary standards and principles that, if applied, assists that potential environmental and social problems allied with aquaculture development are duly attended and that aquaculture grows in a sustainable manner.

Although, offering an enabling environment for sustainable development in mariculture is not only the accountability of aquaculture producers and government, but also the responsibility of media, scientists, financial institutions, and special interest groups. Added principles and standards comprise the ICES Code of Conduct, as well as NACA Code.

7.2.2. Certification

Aquaculture operations can be certified as: (i) creating cultured species to codes or guidelines of practice, (ii) producing cultured species to recognized and reputable standards, or (iii) through operational audits and appraisals as producing species to described criteria. The following section discusses these three methods of certification:

- Aquaculture operations are legitimately certified as producing cultured species in accordance with the codes or guidelines of practice, sometimes followed by Eco-labeling of the product. For instance, the global aquaculture alliance (GAA) is an international, non-profit trade association, that is responsible for promoting environmentally reliable aquaculture through an Eco-labeling program known as "Responsible Aquaculture Program," that comprises codes of conduct for responsible certification and aquaculture production standards.

  There are other schemes that are aimed at giving more emphasis on third-party certification. As an instance, the GAA Codes of Practice for Responsible Shrimp Farming comprises sections on mangroves; design and construction; site evaluation; feeds and feed use, therapeutic agents, and other chemicals; shrimp health management; general pond operations; community and employee relations; and effluents and solid wastes.
- Aquaculture operations may be specialized as producing cultured species to recognized and reputable organic standards. For instance, the International Federation of Organic Agriculture Movements Basic Standards (IFOAM) offers organic production standards for aquaculture as well as agriculture that are used by certifying bodies and standard-setting organizations across the world as a framework for development of certification criteria.

IFOAM comprises criteria for rearing of fish and servicing of cages; feeding; water quality; health; fish re-stocking, propagation of fish stocks and breeding; breeding and origin; and transport, killing, and processing.

Some organizations that are using IFOAM standards are KRAV Kontroll AB Organic Standards 1999 (certifies salmon, trout, Arctic char and Brown trout farming in Sweden), The Naturland Standards for the Production of Salmon and Other Cold Water Fish (primarily in use in Germany and Ireland for trout and salmon farming), National Association for Sustainable Agriculture Australia (used in Australia, Sri Lanka, PNG, and Indonesia),

AgroEco (based in Holland, organic shrimp farming in Ecuador); and BioGro New Zealand Production Standards.

- Aquaculture operations may be certified, through operational assessments as well as audits, as producing cultured species to labelled criteria. Certification is followed by ecolabelling of the product and usually needs the implementation of a documented environmental management system (EMS).

The International Organization for Standardization (ISO) has advanced sets of generic management system standards that can assist in general standards and criteria for the advance of an EMS. The ISO 14001 EMS has been taken into use by several organizations as a basis for environmental certification. One such organization is the Audit Scheme (EMAS) and European Eco-Management.

EMAS is a management tool for companies as well as other organizations to assess, report, and advance their environmental performance. Participation is voluntary and prolongs to private or public organizations operating in the European Economic Area and European Union.

An increasing number of candidate countries are also applying the scheme in planning for their agreement to the EU. Some companies, local authorities as well as other independent organizations outside the European Economic Area are already putting EMAS into practice unofficially and profiting from continuous amended environmental performance.

It is generally seen that adequate monitoring programs are critical for maintaining and attaining an environmentally friendly mariculture industry. GESAMP has fabricated a working definition for monitoring with respect to aquaculture as "the regular collection, normally under regulatory mandate, of biological, chemical or physical data from predefined locations such that ecological changes attributable to aquaculture wastes can be quantified and evaluated."

GESAMP also highlight that in a way to have effectual regulatory tools, monitoring programs should be integrated with simulation models that can be help in predicting the impact of a given operation and reply with corrective action if the threshold levels for environmentally suitable impact are broken. Although, there is no clear definition of what such environmentally satisfactory impact means, and what indeed are safe boundaries for impacts.

Monitoring and regulating the production process and the degree of the operation is also a requirement to assimilating mariculture into coastal zone planning. It is only when passable data are existing in place that environmental, including biodiversity, as well as mariculture needs can be firmly articulated.

It thus follows that integration will be fruitful when all participants (users of the coastal resource) are able to recognize their environmental needs and impacts while representing a high level of credibility in their valuations. To enhance public confidence, it is suggested that the results of ongoing monitoring programs are reachable to the public.

Monitoring programs should focus on the critical impacts of mariculture. It has been recommended that the following criteria should be used to handpicked the impacts on which to place the main emphasis:

- The amount of the impacts must have significance for both the mariculture and the environment operation;
- The impact must be suitable for supervising, such as, routine analytical methods should be available and the signals must be different for background levels;
- Scientific information should be available to set passable EQS;
- The monitoring should be cost effective, as many mariculture processes are small enterprises.

7.2.3. Aquaculture Laws and Regulations

Because of surging global concerns over the environmental impact of aquaculture and its impacts on biodiversity, several countries have enacted laws that particularly control the aquaculture industry. Unluckily, several countries are still dependent on more general environmental protection laws or local environmental plans that occasionally challenging to enforce with respect to aquaculture operations or are susceptible to legal or political manipulation.

Several countries need EIAs to be conveyed out for planned aquaculture projects, leaving traditional operations unfettered.

Some instances of aquaculture-related laws and regulations comprises:

- To date most EU Member States, have firm that only finfish aquaculture projects should be focus to EIA and few shellfish projects, irrespective of scale, have been subject to EIA. The MARAQUA project suggested, although, the adoption of the EIA process for all aquaculture operations;
- There are considerable cost implications in conducting out a full

EIA for new developments, specifically for smaller operators that dictate the industry in several countries. Such a situation may be fixed somewhat by implementing a "scaled down" version of an EIA for proposed developments. As an alternate to a full EIA, Fernandes et al., 2000 recommended that in some cases environmental studies of a more restricted nature could be conveyed out and the results delivered to the regulatory authorities in the form of an "Environmental Report" when making an application for a fish farming permit.

- ICES (2002) decided that there is a necessity for the modification of criteria to resolute if an EIS is needed for shellfish operation and suggested that further work should be conveyed out to find out appropriate threshold criteria to decide which sort of shellfish projects should be subject to: (i) an EIA; (ii) and environmental report; or (iii) no environmental assessment;
- In Malaysia, an EIA is needed for "land-based aquaculture projects convoyed by clearing of mangrove swamp forests covering an area of 50 hectares or more," pursuant to the environmental quality (prescribed activities) (EIA) order 1987 of the Environmental Quality Act 1974. Although, the EIA law with respect to aquaculture projects is weak. There is a voluntary "Code of Practice for Aquaculture" (6th September 1999);
- New South Wales (Australia) governs aquaculture pursuant to Part 6 of the Fisheries Management Act, the Fisheries Management Aquaculture Regulations (1995) and by the Environmental Planning and Assessment Act. These laws label aquaculture as a "designated development," that needs an EIA;
- In 1996 the Supreme Court of India ordered the closure and rehabilitation of numerous nontraditional large-scale aquaculture industries that have instigated harm to India's coastal areas. Conventional approaches and better traditional methods of aquaculture were allowed to endure.

The Supreme Court based its order on the cognitive that aquaculture is an industry and is hence forbidden under the Coastal Regulation Zone (CRZ) notification (which excludes new industries or extension within the CRZ area); In Chile, an EIA could be needed for any project planned in coastal areas pursuant to the Environmental Framework Law (No. 19.300);

Sri Lanka's National Environmental Act (NEA) needs an EIA for any fisheries project that is larger than 4 hectares and forbids any person from discharging, putting, or emitting waste into the environment that result in causing pollution without a license issued by the authority or in agreement with standards prescribed under the act.

This comprises the issuing of license by the Central Environmental Authority (CEA). Moreover, the Fisheries & Aquaculture Resources Act No. 2 of 1996, comprises rules for the management, conservation, regulation, and development of fisheries and aquatic resources in Sri Lanka.

> In last few years, the growing global concern for the destruction of mangrove habitat by aquaculture operations, most usually shrimp farming in brackish-water ponds, has led to more severe regulations.

In 2000, only Ecuador and Belize had laws in place that particularly forbid the destruction of mangroves resultant from aquaculture projects, while other countries, such as Malaysia, Costa Rica, India, China, and Thailand relied on EIAs or moratoriums on development projects in mangroves, and general environmental regulations.

Other legislation is in place that should be cogitated while evaluating proposals for mariculture projects. In the European Union, the Birds Directive (79/409/EEC) and the Habitats Directive (92/43/EEC) are significant in this respect.

7.2.4. Specific Criteria on the Effects of Mariculture on Biodiversity

Although, it was noted that the Group was unable to determine any established sets or reviews of mariculture criteria, some instances of specific criteria that have been approved as indicators of environmentally comprehensive or sustainable aquaculture can be cited.

For instance, the Government of Japan has endorsed regulations to ensure sustainable aquaculture, with respect to improvement and protection of aquaculture grounds and the spread of disease.

In the case of surface aquaculture, it was firm that oxygen levels in cages should be more than 4.0 ml/L of sea water, the amount of sulfide in the mud under cages should be fewer than the oxygen exists to reset the benthos and sulfide, such as lugworms, should be existing in the mud under the cages.

Norway has founded quality criteria for fish oil used in fish feed by commercial fish farmers. The feed should include 5% free fatty acids, have a total oxidation value of 30 and contain only 0.5% water and impurities. This is imperative because increasing the fat content in feeds assists in minimizing wastage by supplying a ready energy source and curtailing the nitrogenous waste producing breakdown of protein for energy.

7.3. AQUACULTURE REGULATORY FRAMEWORKS

The development of sustainable aquaculture operations relies on the establishment of several essential institutional as well as technical preconditions. Public sector institutions play a vital role in development of an aquaculture project at each stage.

It is important to address a number of problems in order to gain the full potential of every aquaculture operation that occurs and minimizing the risks and impacts at the same time. Careful planning, zoning as well as prioritization of sites among the different potential users are the basis for a sustainable aquaculture operation.

Conflicts with other economic activities (for example, tourism, agriculture, forestry, fisheries, and subsistence users) and other users can be expensive and result in the failure of the investment as well as the irreversible damage to economically and ecologically critical resources.

The clear identification of national procedures and policies is another key factor. Before the development of the industry explodes, governments should establish clear procedures as well as policies. Such procedures and policies should address the most common issues and impacts that are involved in aquaculture activities and do not have to be exhaustive. In simple terms, the procedure for obtaining licenses, concessions, permits, land titles and approvals for projects should be transparent, clear, and simple.

This is one way to ensure that government officials have clear and transparent guidance about how to proceed with approvals and that proponents comply with the appropriate procedures. A sound legal framework should be implemented to ensure the rights of aquaculture farmers in addition to protect the interests of the state. This framework should involve least possible requirements for the institutional set-up as well as procedures stated above.

Other likely methods to regulate aquaculture activities involve land-use classification and zoning, bans, mitigation plans, restrictions, user fees, EIA, permits, performance bonds, and monitoring requirements. Specific methods applied widely to regulate aquaculture involve effluent discharge permits, EIAs, restrictions on feed use, limitations on the use of non-native species, standards for feed composition as well as other management practices.

7.3.1. Evolution of Aquaculture Regulatory Framework

A legal framework has been built up in countries where development of aquaculture has occurred that allows for control of the operation of and access to aquaculture activities.

The legal framework also establishes the institutional framework as well as orientations for the management of activities related to aquaculture that includes formalities concerned with specific procedures. Other sectorial laws such as the land law, including the use of public domains as mangrove areas, environmental law, the water law, animal health and animal disease law, fisheries law as well as trade law directly influence aquaculture.

A number of the concerns as well as issues may be regulated under a more general legislative regime and involved are not unique to aquaculture.

Moreover, several regulations as well as laws in place may not even applicable to aquaculture directly. Therefore, are often applicable in an inconsistent manner to the sector. Conflicts may take place among the agencies and institutes involved or within the range of legislation applicable to aquaculture.

A number of nations have framed particular rules and regulations regarding aquaculture within aquaculture specific legislative text, basic water law, fisheries law, or any other piece of legislation. These laws have a tendency to organize a few fundamental principles related to aquaculture and later on invest the legitimate authority with the supremacy to control aquaculture.

Even though aquaculture as an activity is closer to agriculture as compared to capture fisheries, it has exclusively become common to control aquaculture and capture fisheries in the similar piece of legislation. Governments find it useful to have the same authority enforce and control both sectors in countries where the aquaculture sector is being developed.

Nevertheless, both should be distinguished institutionally as well as legally. From a legal school of thought, current fisheries legislation generally does not form a satisfactory ground for bringing aquaculture into play. Developing and adopting a new law is a smaller process while changing a law is often considered as a bigger process.

There are several concerns that are required to be taken into consideration for aquaculture that is not obviously provided for in fisheries legislation like disease management and control, management of temperature and water quality, quality of feeds as well as use of drugs and chemicals. Therefore, the soundest and best regulation of aquaculture is done by individual legislation.

Recently a number of fisheries laws adopted contend with particular aspects related to aquaculture. The aquaculture and fisheries sectors are even mentioned side by side in some cases, which indicate that the legislature gives same significance to both sectors.

Instances involve Fisheries and Aquaculture of Madagascar Ordinance of 1993, Malawi's Fisheries Act of 1997, Burkina Faso's Forestry Code of 1997, Chad's Forestry, Fauna as well as Aquatic Resources Act 2008 and Zambia's Fisheries Act 2011. Environmental damage is caused by the expansion of the aquaculture industry.

Moreover, governments are increasingly accepting the fact that inappropriate and uncertain legislative arrangements are hindering the sustainable development of the sector up to a great extent. Laws and regulations related to aquaculture have developed some commonality in respect to approach as well as the required minimum elements for their comprehensive management.

This is being driven by number of factors, involving: evidence of social disruption as well as environmental damage as a result of the rapid and largely unregulated expansion of aquaculture activities; a growing emphasis on improving the safety as well as quality of aquaculture products in international trade; greater political attention as the economic importance and potential of aquaculture become more apparent; and greater awareness that institutional arrangements an inappropriate laws can significantly constrain the development of the sector.

In recent years, the following analysis highlights the most common elements found in aquaculture legislation and the most significant trends as well as initiatives in national aquaculture legislation. The requirement to develop a legal framework in addition to policies for aquaculture is now widely recognized. Governments are required to magnify institutional and legal instruments to:

- Upgrade food safety and quality to safeguard consumers and meet the standards of importers;
- Integrate aquaculture concerns into resource use as well as development planning;
- Upgrade the management of aquaculture, specially where it has the potential to be socially or environmentally unsustainable; and
- Identify aquaculture as a separate agricultural sector. It is considered that issues like planning and management, registration,

and access in addition to the numerous environmental impacts of aquaculture should be dealt in a consistent manner to protect the environment, consumers, the industry, and other resource users.

Countries that have become farmed-shrimp producers, such as the Philippines have recognized the need to incorporate these aspects into new legal frameworks.

The Philippines Fisheries Code of 1998 provides for the management, conservation, and development of aquatic as well as fisheries resources, and involves a variety of issues like leasing of fishponds, incentives, and disincentives for sustainable aquaculture practices, the use of public lands, the establishment of an aquaculture Code of Practice, and the establishment of an Aquaculture Investment Fund.

7.3.2. Establishing an Institutional Framework

Planning and administering expansion of aquaculture in a sustainable manner is the most important role for national governments. Setting the national agenda for aquaculture development involving the necessary regulatory, administrative, and enforcement mechanisms for its realization is the exclusive role of the government.

Planning for aqua cultural policy in addition to comprehensive public-sector development has become more widespread. Based on financial, social, and environmental sustainability, governments should develop clear as well as achievable policies for aquaculture development in order to foster sustainability.

In addition, these policies based on sustainability should be elaborated using dialogue among the private sector, non-governmental organizations (NGOs), international organizations and national agencies. As stated in Article 9.1.1 of the FAO (1995) CCRF, the need for states to establish, maintain, and develop a suitable administrative and legal framework for the development of aquaculture carries with it a variety of institutional responsibilities.

Legislation is only fruitful when it is implemented and enforced in an appropriate way and ignorance of these matters could result in fatal of the realization of the policy objectives which regulation aims to attain. The nature of aquaculture involves a variety of overall supervisory functions that are require to be exercised by governments over the functioning as well as general policy of the industry.

Questions related to national concern, like the formulation of a wide-ranging national policy, require being the accountability of the central government in order to secure a comprehensible overall direction for the aquaculture industry. In the same way, the enactment of suitable laws to administer aquaculture must rest with the central legislative body in addition to the government.

Nevertheless, this is unable to preclude the enactment of executing regulations at local, regional, and national levels. The allocation of overall responsibility for aquaculture legislation and policy to the central government is included in the general pattern for the distribution of administrative responsibilities.

Nevertheless, this responsibility is delegated to diverse degrees among local bodies in federal states, each one of which is liable for local as well as technical concerns. Administrative overlap and interference are the most common problem in the administration of aquaculture policy. Conflicts are frequent and efforts are duplicated between different levels of government as well as between government departments.

7.3.3. Environmental Management in Aquaculture

Uncontrolled and unregulated aqua cultural development can contaminate the environment, disturb rural communities, and support overuse of natural resources. The objective of environmental management is to lessen, mitigate or avert adverse environmental effects of human activities to allow more sustainable uses of resources through improved practices (Figure 7.3).

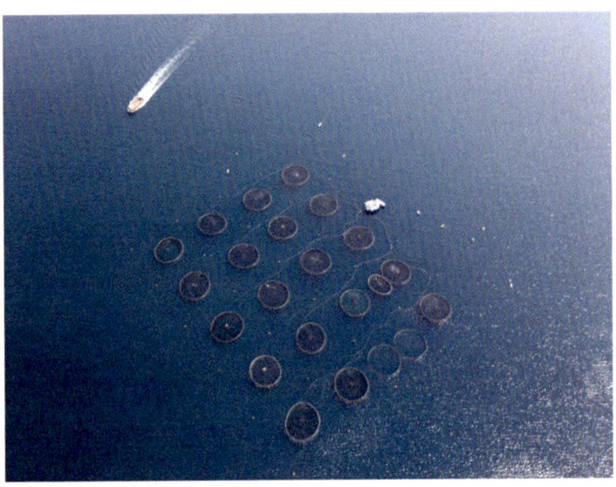

Figure 7.3. Environmental management in aquaculture.

Source: Image by Civil Eats.

The following conditions can be valuable to decide whether environmental management is effective: (i) the possible opposing environmental impacts of human activities must be recognized; (ii) standards must be prepared to stipulate changes in variables that are allowable without causing unacceptable environmental effects; and (iii) technology-based management practices must be accessible to avert excessive changes in environmental variables.

Government and the private sector conventionally share the accountability of ensuring environmental quality. Government is accountable for instituting standards for environmental management, and accepting regulations and laws to impose compliance with these standards.

The private sector is accountable for the cost of evaluations to decide probable ecological impacts, but governments should offer provide guidelines and accept the conclusions and suggestions of such evaluations. As aquaculture activities develop, environmental aspects become of a growing concern.

Management accidents or deficiencies at aquaculture facilities can have noteworthy negative impacts. Potential effects of aquaculture activities comprise lower water quality, negative impacts on the ecosystem, and declines in other economic activities. Environmental concerns form part of licensing procedures, from the probable environmental impacts of establishing an aquaculture activity to the influences that arise through the real operation of an aquaculture farm.

Thus, proposed legislation should subject suggestions to an EIA when the operation hovers grave environmental harm. A method that is getting growing consideration in aquaculture management is the ecosystem approach to aquaculture (EAA).

The objective of the EAA is to endorse the sustainable development of linked ecosystems and manage the aquaculture sector in a way that does not endanger the wider ecosystem. Environmental considerations are extremely imperative to attain these objectives.

The ecosystem approach should chiefly be stated in a country's policy framework. Legislation under the ecosystem method can be done in various ways, and is usually done through a series of provisions. The EAA can be seen more usually as an overriding principle, and thus be placed in the preliminary articles of an aquaculture act. Legislation under the EAA also comprises provisions that intensify institutions and allied management systems.

7.3.3.1. Site Selection and Environmental Impact Assessment (EIA)

It is generally seen that one of the major reasons for the failure of fish farms or for adverse environmental impacts is the farms locations on low-grade or marginal sites. Thus, a comprehensive site assessment should be conducted to define if site faces are apt for the construction and sustainable operation of a farm. These evaluations should decide the obtainability and water quality, freshwater flow, climatic conditions, as well as other related factors.

The information attained should also be comprised in a thorough EIA. Submission of an EIA is generally a precondition for the issuance of a general aquaculture license. The EIA addresses several factors such as the projected size of the aquaculture farm, and the area's sensitivity involved.

As the location of an aquaculture farm at an unwanted site can be catastrophic, the suitability of the site is one factor that should be particularly given greater emphasis by the administration prior to the grant of a permit. This decision on a permit application also lets government agencies to assimilate the siting of aquaculture farms with plans for the management of lakeshores and coastal zones.

In some of the instances, the projected location of a planned aquaculture operation should be inspected along with the other environmental effects of the proposal. It is thus imperative to when an EIA is needed. When the application for a permit is obtained, an initial decision should be taken as to whether the project will be topic to an EIA.

The legislation should suggest clear criteria for decide whether an evaluation is needed for a planned aquaculture project. Because an EIA can levy great expenditure on the applicant, it should be needed only for those projects that generate a genuine risk of environmental damage.

The standards for deciding whether this risk lasts should center on factors such as the size of the planned project, the highest carrying capacity of the site, probable discharge of waste into sensitive areas, protecting exotic products or species from modern technology or genetically adapted organisms, and whether the project hovers exceptional or endangered species.

When an EIA is needed, the decision of regulators to allow an authorization, and the conditions of the authorization itself, should imitate the results of the assessment. When an EIA is not needed, the regulator should grant permission on the basis of explicitly stated principles.

In a way to draft good aquaculture legislation, "it is essential that specific decisions about the granting or withholding of development are made against a background of explicit policy principles."

7.3.3.2. Effluent Regulation

One of the main concern resultants from the activities of aquaculture farms lies in their capacity to harvest large amounts of waste products which, if not managed properly, possess the ability to sternly cause greater damage on the environment. Waste products can impact the obtaining aquatic environment, and pose a threat to human health.

There are two separate regulatory tactics to the control of unsatisfactory waste from fish farms. The first is the obligation of a discharge licensing scheme that sets greatest quantities of contaminants that may be exists in waste water from fish farms, with an allied sanction when these quantities are surpassed.

The second approach comprises the restriction or prohibition of use of specific chemicals and drugs which are known to have a destructive effect upon the quality of obtaining waters and ecosystems.

Discharge licensing systems that comprise water quality and/or water volume criteria needs continuous monitoring at particular intervals to validate compliance. This sort of discharge licensing is apt for large operations with one or more effluent outfalls. The most effective system appears to use overall discharge permits that necessitates farmers to apply a set of specified operational Best Management Practices (BMPs), with negligible basic monitoring necessities.

7.4. DEVELOPMENT OF SUSTAINABLE METHODS

It is worth noticing that the key to affordable and sustainable aquaculture practices lies in the independence from natural resources through remediation or recycling of fish production wastes. Although, sustainability, and the lessening of ecological impacts relied upon the governance of environmental and infrastructure agencies must improve regulations and legislation across the world (Figure 7.4).

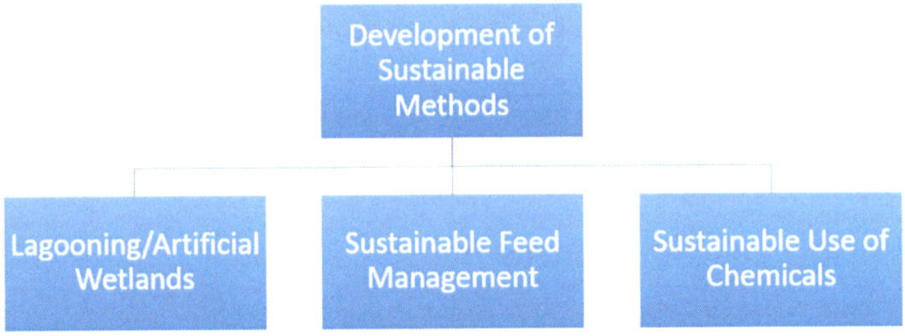

Figure 7.4. Development of sustainable methods.

There are several countries having imperative aquaculture production have some environmental regulations but scarce vibrant frameworks for evolving technologies such as offshore farming. Every farmer should take into consideration formation of sustainable principles, practices as well as methodologies. Thus, it is highly suggested to promote certifications of good practices.

Environmental monitoring costs may be too onerous and expensive, hindering its interpretation or application, and make sustainable development more challenging So that, just representative, reliable, non-redundant, cost-effective, and noteworthy measures should be integrated within environmental monitoring tactics to synchronize production with the ECC of the selected site.

Distinctive methods are being used to lessen the environmental impact of aquaculture practices, but additional research is still required. For instance, the use of selective breeding to lessen the use of chemicals for disease treatment may curtail the population genetic variation and probably enhance the virulence of the pathogenic organisms that may specialize in specific organisms. In addition, the use of vaccines signifies a pertinent method but can cause noteworthy stress for the fish during the vaccination which itself can enhance the disease risk.

7.4.1. Lagooning/Artificial Wetlands

It is generally seen that wastewater lagooning is a highly effective, low-cost solution (initial installation and maintenance is required) for cleansing wastewater from land-based farms (Porrello et al., 2003). The treatment of wastewater mainly comprises of a series of physical, biological, and chemical processes to eradicate contaminants and distinct clean or at least solid waste and reusable water that can be take into use for a number of agricultural or industrial reasons. Such sorts of artificial wetlands are already extensively used for the management of municipal waste and are particularly effective at eliminating excess nitrogen and amassing excess phosphorus in the soil (Vymazal, 2010; De Lange et al., 2013; Almuktar et al., 2018).

This type of Phyto treatment has been tested already with wastewater from fish ponds as means of algal ponds and wetlands and has exhibited to be an efficient system by lessening nutrient contents and modifying physicochemical parameters of water (Porrello et al., 2003; Adeoye et al., 2009; Omitoyin et al., 2017).

Although, lagooning systems needs large surface areas, thus, competing for land space with other sectors. The fitting of aeration systems makes artificial wetlands more effective by augmenting the oxidation rate of pollutants and allows lessening the space required (El-Kamah et al., 2011; Almuktar et al., 2018).

Using the area for extra production makes this method more lucrative. Several halophytes in distinctive types of constructed and natural wetlands are already fruitful used as down-stream filters of industrial, municipal, or agricultural wastewater (Verhoeven and Meuleman, 1999; Vymazal, 2010).

Their use can result in having better economic benefits in comparison to conventional water treatment and they are being examined as possible extractive species in integrated aquaculture as they are able to apply high levels of ammonia that may be toxic to other plants (Glenn et al., 1991; Cardoch et al., 2000; Kudo and Fujiyama, 2010).

Distinctive species of halophytes have been successfully cultivated in the effluents of European sea bass (*Dicentrarchus labrax*) in RAS (Waller et al., 2015) and can be assimilated in artificial wetlands or cultivated hydroponically (Singh et al., 2014). While still carries a tag of a niche market, they are gradually gaining popularity as a delicacy vegetable (Custódio et al., 2017).

Ensuring the integrity as well as function of ecosystems is imperative for the sustainable spread of aquaculture operations. An instance is the practice of Integrated or Silvofishery Mangrove Aquaculture that upholds a definite degree of tree cover in and around aquaculture ponds made in mangrove areas (Budihastuti et al., 2012; Bosma et al., 2016).

Mangroves are imperative tropical ecosystems that offers a nursery and habitat to many species, stop sedimentation of seagrass and reefs meadows, protect coasts from flooding and storm surges, and accumulation of atmospheric carbon dioxide (Nagelkerken et al., 2000, 2008). Several mangrove areas across the would have already been lost to the transformation of coastal habitats, involving for aquaculture production (Valiela et al., 2001; Alongi, 2002).

Building aquaculture ponds while upholding a level of tree cover in and around the ponds is a strategy how to combine food production with mangrove protection (Yunus et al., 2015). Sustainable aquaculture is considered to be one of the imperative ways to safeguard these ecosystems, reserve their biodiversity and secure their sustained contribution to climate protection (Primavera, 2005; Ahmed et al., 2016).

Such a tactic to integrated cultivation is especially significant in the context of tropical developing countries where the significance of aquaculture goes beyond intensive production of high-value species and comprises its imperative contribution to food production as well as livelihoods in rural areas (Harrison, 1996; Edwards, 2000).

Particularly in Asia, low-input, integrated fish farming systems have been a long custom (Chen et al., 2019). Large, government driven projects have minimal success rates if they make dependencies and have stumpy adaptation rates by stakeholders (White et al., 2005). Community-based projects put local actors in the spotlight and aimed at improving the livelihoods through capacity building, business development, and diversification (Pomeroy, 2012; Yuerlita et al., 2013). Extensive, co-managed aquaculture operations need low input and technologies while still being imperative sources of income (Partelow et al., 2018; Senff et al., 2018).

7.4.2. Sustainable Feed Management

Sourcing of aquaculture feed is one of the sustainability cores confronts of marine finfish aquaculture (Klinger and Naylor, 2012). Exaggerated production and the cultivation of high value carnivorous fish mainly relied upon the use of fish oil and fish meal as the main feed ingredients, making it a consumer of seize fisheries products, particularly of nontargeted fisheries and small forage fish (Figure 7.5) (Cao et al., 2015).

This has resulted in causing severe environmental and economic concerns, with feed costs being a large part of total output costs, and vital progress has been made towards sustainability by augmenting feed efficiency (Verdal et al., 2017), turning fish offal into useful silage (Cunha et al., 2019; Gonçalves et al., 2019) or designing plant-based, insect-based, and polychaete-based protein feeds (Boyd, 2015; Pahlow et al., 2015; Rhodes et al., 2016; Gómez et al., 2019; Llagostera et al., 2019).

Figure 7.5. Ensuring sustainable aquaculture.

Source: Image by Worldwide aquaculture.

The challenge has been to replace fish oil with other substitutes and ensure the high proportion of greater unsaturated fatty acids within the feed to sustain the nutritional quality of the fish for human consumption (Little et al., 2016).

While the use of land-based feed may lessen the burden on fisheries, it can considerably enhance the burden on freshwater resources (water footprint), because of water consumption and pollution in crop production (Pahlow et al., 2015).

Although, in the recent some time, several novel approaches have been come into place to lessen excess feed used and loss of food. In rigorous fish farming where feeding is taking place by an automatic system, it is significant to supervise the feeding activity of the fish and regulate the amount of feed to the feeding behavior. Such monitoring can be prepared by using, for instance, an underwater camera technology or other similar tactics that detect uneaten feed and hindering the feeding process (Zhou et al., 2017).

More sustainable feed management can be attained through IMTA in a number of ways. Dissolved nitrogen in wastewater helps in promoting the growth of macro or micro-algae substitutes the need for growth media or fertilizer. These extractive species can in turn be cultivated as a feed source, aggregating their practical value. The macroalgae *Ulva lactuca* that has lower economic value when sold directly, offers high quality feed for urchins or abalone (Neori et al., 2000) and microalgae serve as feed for mussels (Shpigel et al., 1993).

Thus, algae production can be deliberated as a more sustainable industry than enduring to harvest fish for fishmeal. It is projected that if microalgae were used as fishmeal replacement, the effect would be to remove 30% of the fishing pressure thus, helping to conserve marine ecosystems (Beal et al., 2018). Algae may replace fish oil, be feasible and even progress growth of farmed fish so it should be pondered to reduce harvesting fish for fishmeal (Sarker et al., 2020).

Effluents can be directly valorized by producing primary producers (Milhazes-Cunha and Otero, 2017; Wei et al., 2017; Li et al., 2019). Sludges and effluents from aquaculture actions may also be bioremediated by detritivores, decomposers, and biofilms, whose biomass in turn offers useful resource for feed production (Martinez-Porchas et al., 2014; Barnharst et al., 2018; Gómez et al., 2019; Robinson et al., 2019).

Solid wastes are an adequate feed for polychaetes that provides protein and vital fatty acids as feed for fish, crab, or shrimp production, eliminating or reducing the requirement for manufactured feed based on farmed crops or fish meal (Brown et al., 2011; Alava et al., 2017; Pajand et al., 2017).

Fish feed can also be complemented or even replaced through the application of biofloc technology. In this method, microbial growth in the water column of fish ponds or tanks is stimulated through the add-on of carbohydrates in the form of sugar, cellulose, or starch (Avnimelech, 2012).

Heterotrophic bacteria, using these as a substrate to form proteins for growth, need nitrogen, which they take from the surrounding water. Dissolved nitrogen species in the water are, thus, converted into bacterial biomass, accumulating into bioflocs that can be consumed by fish or shrimp (Avnimelech, 2012).

Biofloc technology in tilapia ponds can provide 50% of the protein consumed by the fish (Avnimelech, 2007) and when used as a protein source in shrimp feed, microbial floc meal displayed the similar results or even outperformed fishmeal or soybean ingredients (Kuhn et al., 2009, 2010;

Bauer et al., 2012). Biofloc systems have been largely used in freshwater ecosystem and for herbivorous organisms but new studies comprised pacific and mullets shrimp (Legarda et al., 2019).

The successful application of biofloc technology, though, needs knowledge of the system, close examining to maintain suitable C:N ratios and microbial densities and upscaling from laboratory to economic application (Kuhn et al., 2010; Hende et al., 2014).

7.4.3. Sustainable Use of Chemicals

It is well noticing that there is an increasing tendency to develop methods with the objective of reducing broad chemical substance use, and, thus, reducing environmental pollution. Such substitute methods can have other positive effects on production such as cost reduction for the producer and augmented consumer acceptance.

Methods that have been used differ between numerous factors such as the fish species and disease in question as well as the formation methods used and sorts of the area where the facility is positioned. They can be alienated as precautionary methods (e.g., limiting fish density), methods worrying disease prevention (e.g., herbal medicine and immunostimulant feeds additives), co-culturing of distinctive marine species which both advantages from such a production model, distinctive physical techniques used such as so called "lice skirts," plankton nets around the fish cage, to evade parasite infestations (Stien et al., 2018) as well as other changing from above mentioned aspects (He et al., 2016; Haugland et al., 2017; Dawood et al., 2018; Jahangiri and Esteban, 2018; Escobar-Dodero et al., 2019).

The major aquaculture producing countries have been using guidelines to report disease incidence and to confine the fish density in accordance with those measures because high density of fish or cages greatly enhance the risk of disease transmission, e.g., infectious pancreatic necrosis or infectious salmon anemia (Escobar-Dodero et al., 2019).

Other methods that have been used comprise functional feed additives, which can be managed via feed (He et al., 2016). They are non-nutritive ingredients that impact fish performance concerning, for instance, feed utilization, growth rate, and survival (Jahangiri and Esteban, 2018).

These additives endorse fish growth and can substitute antibiotics in those countries where it is still used as averting and growth promotional measures, and thus evading associated negative effects (Done et al., 2015). Most widely studied options comprise prebiotics, probiotics, symbiotic, plant extracts, acidifiers, nucleotides, and such immunostimulants as lactoferrin and β-glucan (Dawood et al., 2018).

Chinese herbal medicines can be taken into use as an alternative in prevention of disease in aquaculture because of antibacterial, immunostimulant, antiviral, and growth promoting substances (He et al., 2016).

Co-culturing of distinctive fish species can be used to confine pathogens at the time of production. An instance of this could be the use of cleaner fishes, which can restrict the spread of ectoparasites such as salmon lice (*Lepeophtheirus salmonis*) due to the predation by those fish species towards these small crustaceans.

Although, this suggests that adequate health and welfare conditions should be offered for both species, as farming together distinctive species would also create a risk of common disease spread, e.g., amoebic gill disease when lumpfish (*Cyclopterus lumpus*) are cultivated together with Atlantic salmon (Haugland et al., 2017).

Genetic improvement through selective breeding has been widely used in aquaculture to enhance growth performance of fish. In Europe, around 80% of aquaculture initiates from selective breeding largely because of this purpose. Although, it is probable to select fish for other traits such as disease vulnerability (Yáñez et al., 2014).

In addition, vaccines are being used to curtail the requirement for antibiotic treatments. Although, the method can be allied with some negative effects such as strain caused to fish at the time of treatment when handling of fish is indulged (Miccoli et al., 2019).

Lytic bacteriophages can be used as therapeutic agents against marine bacterial diseases such as those allied with vibrio. Although, more research is still required to apply these viruses under field conditions and to evade dispersal of unwanted genes and effects on fish microbiota (Kalatzis et al., 2018; Plaza et al., 2018).

7.5. CONCLUSION

Until the 1990s, aquaculture was seldom stated in any enforcement provision. When aquaculture was dealt with under a fundamental fisheries law, the law enforcement sections had regularly been drafted with only capture fisheries in mind. Progress has surely been made in newly adopted fisheries and aquaculture laws. In several countries, practicing aquaculture without approval, discharging wastewater from aquaculture facilities without an authorization or importing aquatic organisms or fish without approval all result in some kind of described penalty.

Overall, the enforcement systems that have been implemented mirror conventional sanctions schemes, involving administrative measures (such as the suspension or revocation of licenses) and fines. Moreover, one often finds provisions defining the duties, rights, and responsibilities of enforcement officers, such as the right to enter, review, and search aquaculture facilities. Although, because of the overlap of laws and institutions indulged in aquaculture, the implementation of enforcement mechanisms remains challenging.

One of the key problems, specifically in developing countries, is restricted budgets and insufficient staff that suggests that aquaculture rules often cannot be enforced properly. Thus, options other than the classic "command and control" mechanisms are being discovered in a way to encourage farmers to make more effective use of resources and to take full accountability to minimizing or mitigating environmental variations caused by their aquaculture activities.

REFERENCES

1. Ap, D., (2021). *Regulatory Framework for Mariculture Development and Management.* [online] www.researchgate.net. Available at: https://www.researchgate.net/publication/262916171_Regulatory_Framework_for_Mariculture_Development_and_Management (accessed 10 September 2021).
2. Carballeira, B. C., Cerbule, K., Senff, P., & Stolz, I., (2021). Towards environmental sustainability in marine finfish aquaculture. *Frontiers in Marine Science*, 8.
3. Ocean Health Index, (2021). *Mariculture: Sustainability and Regulations.* [online] Available at: http://www.oceanhealthindex.org/methodology/components/mariculture-sustainability-and-regulations (accessed 10 September 2021).
4. Takoukam, P., & Erikstein, K., (2013). *Aquaculture Regulatory Frameworks.* [eBook] Available at: http://www.fao.org/3/bb124e/bb124e.pdf (accessed 10 September 2021).
5. Zedan, H., (2004). *Solutions for Sustainable Mariculture-Avoiding the Adverse Effects of Mariculture on Biological Diversity.* [eBook] Available at: https://www.cbd.int/doc/publications/cbd-ts-12.pdf (accessed 10 September 2021).

Chapter 8

The Future of Mariculture

CONTENTS

The idea of the mariculture revolves around the idea of saving environment as whole. This chapter discusses about the various food which is needed for the human population growth and for the fish. Along with this how the cultivation of fish is being done according to the different regions. Moreover, there are some new startups who have explained the new ways of feeding the fish. In addition to it, the role of the integrated mariculture in future aquaculture is being discussed in this chapter and how different type of fish or the marine life of food has been discussed.

8.1. INTRODUCTION

In the coming decades for the nutritious food the human population growth, rising incomes and preference shifts will considerably increase the global demand. In many countries malnutrition and hunger is still a plague. And by 2050 projections of population and income suggests a future need for more than 500 megatons (Mt) of meat per year for human consumption.

It has been said that scaling up the production of land-derived food crops is challenging and because of declining yield rates and competition for scarce land and water resources. In the food security and global supply land-derived seafood has an important role but its expansion is also constrained.

In the substantial environmental externalities that affect the water the expansion of land-based aquaculture has resulted in this way, similar to the other land-based production. It also has effect on the soil, biodiversity, and climate, and which compromise the ability of the environment to produce food.

In the seafood production despite importance of terrestrial aquaculture in seafood production, there are many countries who have restricted the use of land and public waters for this purpose, which constrains expansion such as China, notably the inland-aquaculture produce.

Though, for the food security inland capture fisheries are important and by the ecosystem constraints their contribution to total global seafood production is limited and expansion is hampered. Therefore, to recognize that land-based sources of fish and other foods are also part of the solution and in order to meet the future needs, it has been said that whether the sustainable production of food from the sea has an important role in future supply.

From the wild fisheries and species food from the sea is being produced, farmed in the ocean (mariculture), and currently accounts for 17% of the global production of edible meat. Moreover, the supplementary information is available related to the protein, as in land-based foods the sea contains bioavailable micronutrients and essential fatty acids that are not easily found and thus uniquely poised to contribute to global food and nutrition security.

There are the widely published reports which are based on the climate change, overfishing, pollution, and unsustainable mariculture which give the impression that sustainably increasing the supply of food from the sea is impossible.

On the hand, it has been said that in policies and practices unsustainable practices, regulatory barriers, perverse incentives, and other constraints may be limiting seafood production, and shifts could support both food provisioning and conservation goals.

So, in this study for meeting global food demand in 2050 there is an investigation on the potential of expanding the economically and environmentally sustainable production of food from the sea. This is done by estimating the extent to which food from the sea could plausibly increase under a range of scenarios, including demand scenarios under which land-based fish act as market substitutes.

To global food supply the future contribution of food from the sea will depend on a range of ecological, economic, policy, and technological factors. On the ecological capacity estimates are solely based and are also useful, but do not capture the responses of producers to incentives and do not account for changes in demand, input costs or technology.

There is a construction of global supply curves of food from the sea that explicitly account for economic feasibility and feed constraints, to account for these realities. In the wild fisheries and in mariculture sectors first derive the conceptual pathways through which food could be increased (Figure 8.1).

After this to estimate the sustainable supply of food from each seafood sector at any given price empirically derive the magnitudes of these pathways. In the end, to estimate the likely future production of sustainable seafood at the global level there is a need to match these supply curves with the future demand scenarios.

There is no doubt about this that aquaculture is the marine equivalent of agriculture. It has been said that mariculture is the same thing, but in the open ocean carried out offshore rather than the artificial pools on land where aquaculture is most commonly performed.

Figure 8.1. The future of mariculture.

Source: Image by unsplash.com.

Till now, that took the form of standard floating pens like those seen above, utilizing only the surface of the water and the portion of the water column immediately beneath it. The same thinking that traditional aquaculture employs, just transported out to sea. Therefore, it is important to know that how to utilize the ocean in a three-dimensional fashion, as more innovative companies have been hard at work thinking for this.

There is a Kampachi Farms which is formerly known as the Kona Blue and it operates a cutting edge maricultural operation today off the coast of Hawaii. By the Ocean spar corporation initially they made use of farming enclosures designed and manufactured before migrating over to their own in-house design, a geodesic sphere they call the aqua pod.

The design which is superior will remains to be seen. It has been said that the capacity of the aqua pod is slightly lower, but they are designed to drift freely. Into the prevailing ocean currents, they can in fact be released and then picked up elsewhere for harvest. Along the way the advantage being waste is released and has no chance to become concentrated in any of the particular are.

For the harvest pods are partially surfaced, which confines fish to a smaller volume, making them easier to scoop out. Preferentially the oldest fish are being harvested. The sick is also being released. By making this type of fish farming a net benefit to the environment in at least one way and those that get recover will add to the local population of that species.

Though, surrounding this practice there exist legitimate environmental concerns. Though, by far for more of many to eat fish rather than beef its better and fresh water cost of raising cattle as well as the healthier qualities of fish meat. It has been said that prevent disease from spreading between such densely cohabitating fish there are the issues relating to the use of antibiotics.

Recently it has been seen that there is the well-founded objection, as the fish were fed by the other fish. For the other reasons smaller species considered inedible to humans or undesirable. In order to cultivate the species that people do want to eat this would only deplete those species. So, for this there is a solution that a semi closed loop wherein the fish poop nourishes algae that grows below the pens, then the algae is processed into food for the fish.

There is the problem of scaling up mariculture operations, with most of the environmental issues now either sorted or with a solution imminently within reach. As, it has been seen that to tend to even Kampachi Farms only has a few dozen enclosures. But to rely more and more on maricultural raised fish for protein the overfishing kills off wild fish populations.

In any given operations currently, the few enclosures can be maintained, inspected, and harvested by a small team of scuba who delivered to the work site by motor boat every week. But it has been said that when such operations have hundreds of enclosures, or thousands, that isn't going to cut it. As it has been explained above, that a permanent manned presence on the site is being needed.

With the largest manned maricultural structure aptly named Norwegian company Ocean Farming has leapfrogged ahead of competitors with yet devised. There is an expectation to become this norm, though as vertically stacked enclosures become more widespread to make more efficient use of the plot of ocean, they have a license to utilize, it may become economical to consider including saturation diving capabilities.

As into the water columns everyone extends the farms deeper and deeper, divers will have less and less time at depth due to the increasingly rapid absorption of nitrogen into their blood and soft tissues. To start housing divers in an undersea habitat like Aquarius Reef Base, tending to the 'crops and 'livestock' from below, rather than above, there is an inflection point beyond which it will make the sense.

As far back as the 1970s this eventually necessity was predicted. To a scale where saturation diving would become crucial during the heyday of undersea habitat programs it was anticipated that future oceanic farming operations would grow, if the divers who maintain and harvest the enclosures are to have the bottom time, they need to do their job.

Nowadays, in that case Nemo's Garden gives a glimpse of the future of undersea cultivation, albeit of terrestrial plants. While for the drivers the current state of the art for mariculture is the unmanned monitoring buoy or modest floating accommodations and for those who live and work on-site the future will be something like the maricultural equivalent of an oil rig, or a fully submerged seafloor habitat. If anyone like the farmhouse of the future.

8.2. THE FUTURE OF AQUACULTURE

Only a tiny fraction of the ocean's area could meet the entire world's demand for fish through the aquaculture. By the scientists from UC Santa Barbara, and NOAA, the National Oceanic and Atmospheric Administration (NOAA) this is the form of a study which is being conducted. To meet the growing issue of global food insecurity that is quite a prestigious team of experts and they suggest that aquaculture or what commonly call fish farming has the potential.

To support both human livelihoods and economic growth marine aquaculture provides a means and an opportunity, in addition to providing food security, said co-author Ben Halpern of UCSB. It is not a question of if aquaculture will be part of future food production but, instead, where, and when (Figure 8.2).

Figure 8.2. World's fastest growing food production.

Source: Image by unsplash.com.

8.2.1. The World's Fastest Growing Food Production

Presently the world's fastest-growing food production sector daily Scandinavian Norwegian-seafood-farm, farming salmon, farmed seafood, Norway, fish farming, aquaculture, sustainable seafood salmon Aquaculture. To meets its own domestic seafood needs space suitable for aquaculture is certainly not a limiting factor and every coastal country has the potential.

In the oceans there are only a couple of countries that are producing the vast majority of what is being farmed right now which is being said by the lead author Rebecca Gentry and across the world aquaculture could be spread a lot more. For example, with Chile producing 31%, followed by Scotland and Canada Norway produces 33% of the world's farm raised Atlantic salmon.

8.2.2. In United States

At a cost to the country of over $13 billion, the United States annually imports more than 90% of its seafood and also the farm raised salmon, aquaculture, health magazine. Therefore, through aquaculture to meet its total domestic demand by using only 0.01% of its coastal waters the US has the potential.

Only in the most productive areas if aquaculture were developed, the oceans could theoretically produce the same amount of seafood that the world's wild-caught fisheries currently produce globally, but in less than 1% of the total ocean surface a combined area the size of Lake Michigan.

By the 39% aquaculture is being expected to increase in the next decade, as it is being said by the Froehlich and a researcher at the NCEAS. The amount of biomass aquaculture produces has already surpassed wild seafood catches and beef production buy not only is this growth rate.

8.2.3. Crucial That Fish Farming Be Well-Managed and Monitored

It has been said that it is crucial for industry, science, conservation, and government to collaborate and ensure that fish farming is well managed and monitored and Froehlich added a very important caveat.

It has been seen that like any food system, aquaculture can be done poorly. In the 1990s Froehlich said, in reference to the boom and bust of shrimp farming, which is being resulting from the poor management. To shape the future of food for the betterment of people and the environment this is really an opportunity.

8.2.4. Can Aquaculture Be Made Sustainable and Safe?

By the guardian.com in "Thousands of Atlantic salmon escape from fish farm into Pacific but, there are many potential problems documented with fish farming as reported (Figure 8.3).

Figure 8.3. Aquaculture be made sustainable and safe.

Source: Image by unsplash.com.

8.2.5. The Solar Eclipse

In the Washington State at least that is what they contend contributed to the exceptionally high tides that damaged fish pens and allowed the escape of thousands of Atlantic salmon into Pacific waters. In their nets the local fishermen looking to catch Chinook salmon were finding Atlantic salmon. On the baby salmon nobody wants those fish preying. And nobody wants them getting up into the rivers, as it is said by the Ellie Kinley, whose family has fished in Puget Sound for generations.

For commercial fish farming in Washington State and British Columbia Atlantic salmon are no strangers to west coast aquaculture and are the major species for and this incident is not unprecedented. With the native Pacific species or compete with them for food fishermen and environmentalists expressed concern that the Atlantic salmon would crossbreed. In the late 1990s NOAA estimated that some 600,000 Atlantic salmon escaped into the Pacific.

However, Michael Rust cautioned that these things are kind of couch potatoes and they are also being domesticated. In the Serengeti one should imagine a dairy cow getting lost out, as it does not last very long. However, with this benign view not everyone agrees and the CBC News reports that escaped Atlantic salmon survive and thrive in B.C. waters.

8.2.6. Farmed Salmon Out-Competed by Wild Salmon

Fortunately, in the natural territory of Pacific salmon the history has not shown any negative impacts or invasiveness of Atlantic salmon, as it is being said by the Jeremy Dunn, director of the B.C. Salmon Farmers Association.

In fact, decades back an experiment is being mentioned by him where hundreds of thousands of Atlantic salmons were released into the waters of British Columbia by the Department of Fisheries with the hope that they would become available for sport fishing. For the food they were out competed by the Pacific salmon, as said by the Dunn, and he feels that the same would be the case today.

It has been said that 10 or 15 years ago escapes were much more common, but they do not happen as often today. In the British Columbia the farms are highly engineered and able withstand very fast currents and very rough seas," Dunn said.

8.2.7. Side Effects Inhibit Farmed Salmon

By the University of Melbourne, a recent Australian study was being published which offers another explanation for why escaped salmon cannot compete with their wild cousins. It has been said that farmed salmon are partially deaf, which is possibly a side effect of their accelerate growth. Since to navigate, avoid predators, and find food fish in the wild use their hearing and for surviving hearing-impaired farmed fish are at a distinct disadvantage (Figure 8.4).

Figure 8.4. An illustration of fish cage.

Source: Image by Wikimedia commons.

8.2.8. Introducing Disease

Regardless, of whether the farmed fish survive and breed, experts warn that they still pose risks to wild fish. For the Atlantic Salmon Federation there is a spokesperson Neville Crabbe who said that the bigger concern is that crowded conditions at fish farms are often breeding grounds for diseases and parasites. And now it has been said that there are thousands of possibly diseased fish mingling with wild species across a huge geographic range. It is a disaster and no matter how people try to downplay this.

8.2.9. What Feeds Fish Farms?

To aquaculture is the issue of what to feed the fish and can it be made sustainable is considered as one of the biggest drawbacks such as sardines-pacific sardines, over fishing, over-fishing, fish collapse, schooling fish, seafood. In the 2014 summit Michael Aw has been a Blue Ocean contributor for years and was interviewed.

From the South Africa Michael had just returned where he had documented the sardine run, one of the world's largest marine migrations, that occurs along the coast of South Africa in the Indian Ocean. Michael observed 10 fishing boats, each taking 100–150 tons of sardines per day. In the Norwegian fish the sardines were destined to feed salmon, where it takes nearly 10 pounds of sardines to produce one pound of salmon.

It is estimated that sardines along the East Coast of Africa will be fished out within two to three years if over-fishing continues at this rate, leaving dolphins, sea birds and penguin to go hungry, not to mention the families of local fishermen.

In part of fishmeal and fish oil the feed fed to farmed salmon is made from the fish like anchovies, herrings, and sardines. Fish are being considered less desirable than the salmon they are fed to and normally these are the smaller trash. Though, in South Africa for the local fisherman, the sardines are what he depends on for his family's protein (Figure 8.5).

Figure 8.5. An illustration of fish farming.

Source: image by Wikimedia commons.

8.2.10. Speeding the Growth of Farmed Fish

For the farmed fish there are the great economic activities to grow to maturity quickly. With a higher fat content than their counterparts growing in the wild this has led to their being which is fed by a diet. Unfortunately, to contamination by pollutants like PCBs that accumulate in the food chain this high-energy diet also makes farmed fish more vulnerable.

Around the world since fish feed is sourced from the fisheries, as it can contain pollutants from far distant seas. So, from heavily polluted areas in Asia, Africa or South America salmon grown in the supposedly "pristine waters" of Norway may be fed feed taken.

From marine sources to agricultural or by-product feed the attempts have been made to move away. For example, from the poultry processing in Canada farmed salmon can be fed by-products, but this might lead to further concerns.

8.2.11. Sustainable Farmed Fish

There is more work that needs to be done as there is sustainable farmed fish available. It has been said that the industry needs the innovation to have a truly healthy and sustainable product and to live up to its enormous potential to help feed the world. In the aquaculture industry many players need to reform their processes, take a hard look at the sustainability of their feed sources, and the impact their farms have on the marine environment.

8.2.12. Shrimp Farming

Around the world countless stories describe the impact that shrimp farming has had on mangrove and marine ecosystems. In the follow up article it is such huge and complicated issue that is being addressed which find it here (Figure 8.6).

Figure 8.6. Fish farming in cage.

Source: Image by Wikimedia Commons.

8.3. SEA CHANGE: NEW HORIZONS IN AQUACULTURE

For spawning aquaculture has always been about the potential there is a blue revolution that would overturn the long-reigning paradigm for global food consumption and allow humankind to extend the bounty of seas, lakes, and rivers in fantastic new ways instead of scratching out subsistence in the soil.

But now, aquaculture is all about the need. As, to produce greater results faster than once imagined there is a need for aquaculture and aquaponics.

From the world's growing population, and traditional farming and agricultural methodologies there is a big existential threat that will not be able to scale to meet that need, says Brian Frank, general partner of FTW Ventures, a San Francisco-based firm that invests in innovative food producers.

It has been said that to fix the degradation of the environment there is also obviously a huge movement. In the long term it has been said that the third thing is climate change which will shift how the food is being produced. In aquaculture there is a need to spur innovation, so that people in another quarter-century have access to the nutrients they need but are not degrading the environment.

In the landlocked ponds while raising fish, crustaceans, and mollusks it is being fenced-in ocean pens has gone on for decades, into the horizon the full potential of the industry has kept receding. About 70% of the planet oceans and waterways cover about, but still only about 5% of food production comes from the water. Today, it has been said that the leaders are stepping out to the strong possibility that their industry's time finally has come (Figure 8.7).

Figure 8.7. An illustration of shrimp crustaceans.

Source: Image by Pixabay.

People are still wanting to buy the fish even there is less fish coming out of the sea and catch the quotas which are being reduced., as Jamie Stein, co-founder of United Kingdom-based Devonian Capital, which is investing exclusively in aquaculture startups said this. For the sustainability produced seafood add the consumer preferences and also the environmental impact, along with this one should have a strong marketing story as well as a strong economic story.

As the worldwide fishery outputs aquaculture has picked up the slack which have stagnated due to the scourge of overfishing, restrictions by some governments, and reductions in seafood populations presumably wrought by climate change already. It has been seen that net- and farm-raised seafood have supplied an increase in global per capita fish consumption even with a rising population, as aquaculture has advanced technologically, scaled up, and become more economical. About half of all seafood aquaculture now produces.

In fact, for the human consumption aquaculture in 2015 surpassed wild fisheries as a source of seafood, while a relatively insignificant contributor just a generation ago. By the aquaculture the annual biomass has been produced and now also exceeds that of beef, and its growth rate has outstripped that of poultry production, making it the fastest-growing agricultural sector. It has been seen that for the protein production with the increasing evidence that seafood is an environmentally sustainable method, and aquaculture is likely to keep gaining on terrestrial protein rivals such as poultry.

In 2018 the aquaculture industry reached an estimated $232 billion in revenue, for 80 million metric tons of food, according to the United Nations' Food and Agriculture Organization (FAO), with dozens of species raised in marine, freshwater, and brackish environments.

By 2050 the take is expected to reach $275 billion, according to Adroit Market Research. According to the Adroit with the exhaustion of the ocean's natural fish productivity the demand will be driven by the rising consumption demand for seafood coupled by the uncontrolled exploitation of wild fish over several decades.

8.3.1. Asia Leads the Way

It has been said that the aquacultural production has been dominated by the Asia. In 2016 fully 89% of the output of aquaculture, by quantity, originated in Asia and about 62% took place in China, according to the FAO.

In the three largest aquaculture producers such as China, Indonesia, and India have had to feed large populations, so the region had an early start and because policymakers for this. It has been seen that via strong research funding and friendly policymaking some western countries, most notably Norway and Canada, also have emerged as aquaculture leaders (Figure 8.8).

Figure 8.8. An illustration of mollusks shells.

Source: Image by Pixabay.

In Asia the most aquaculture occurs which helps account for the fact that various types of carp comprise four of the top 10 aquaculture species globally. Tilapia and shrimp have become truly global aquaculture species as well. About 25% of the industry consists of mollusks such as cupped oysters and Japanese carpet shell clams.

8.3.2. Aquaculture Now Produces About Half of All Seafood

In the Salmon much of the today's action in the aquaculture, as because of the fish's high nutritional content and increasing popularity with Western consumers whose popularity as an aquacultural species has been rising. Also, in mechanizing aquaculture and bringing modern technology to the practice the salmon-raising industry has been the leader.

It has been said that to revolutionizing salmon raising, providing new technology for oxygenation, biofiltering of wastes, and the other requirements for raising Recirculating Aquatic Systems (RAS) have been the key. In the fresh water Atlantic salmon to the juvenile stage, after which they are shipped to ocean-based pens.

About anywhere the one key which is RAS can be just put. Over the past several years there has been an explosion of the new projects, including Superior Fresh, currently the largest producer, in Hixton, Wis. Another ambitious initiative is Atlantic Sapphire, a $350 million facility being built in Miami, where its Danish founders envision providing half of America's salmon by 2030.

It has been said that it created even more of a fever behind RAS investments, when COVID-19 hit. In the other industries rich guys who normally would invest now were sitting at home and could not get a piece of salmon, as Terence Barry, an emeritus scientist in aquaculture at the University of Wisconsin-Madison, puts it. After this everyone has released that how much they depend on the foreign fish and how much they need to invest in U.S. aquaculture instead of getting their fish shipped from Norway.

But with raising salmon there are problems. It takes a lot to feed them for one thing and that is one of the main challenges said by the Sophie Ryan, CEO of the Global Salmon Initiative, a consortium of 13 of the largest farmed-salmon producers around the world. It takes out of the ocean the sources of the fish, which could stay in the ocean or be used elsewhere and it can also impact the biodiversity.

To feed the salmon the ratio of fish required previously was three-to-one. But, by the initiative in the large parts, that ratio now has come down to about one-to-one. As salmon production continues to grow by the alternative feed sources including algae, canola, and insects have gained share even as.

It has been said that certification is one of the industry's weapons. For raising salmon there were no universal standards seven years ago. But now, about 60% of global salmon farms are certified and monitored for continuously improving performance in use of antibiotics and farm management, after the creation of the Aquaculture Stewardship Council (ASC).

8.3.3. New Options From Start-Ups

By the challenges to Atlantic salmon farming another dynamic has been created and more startups are offering the species alternatives. Therefore,

Finger Lakes Fish is the one company in which Devonian has invested, better known as LocalCoho, which is growing coho salmon in upstate New York (Figure 8.9).

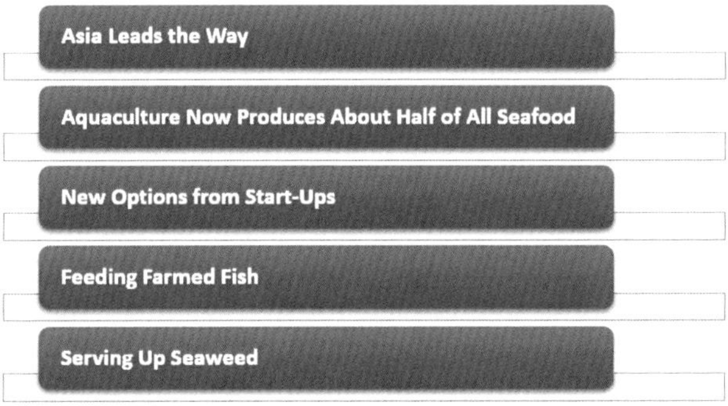

Figure 8.9. Sea change: New horizons in aquaculture.

In the Auburn, N.Y Finger lakes fish has built a 43,000-square-foot RAS facility and plans full capacity for putting out 2,000 fish a week, at an annual run rate of 450 tons. Its two raceways, tanks of oxygenated water where salmon live are six feet deep, 16 feet wide, and 80 feet long. From certified organic eggs and raised to maturity the sushi-grade coho salmon are hatched.

Without the antibiotics or pesticides or contaminants in the water Finger Lakes raises its coho salmon and keeps them in the tanks through the adulthood. With no possibility of diseases or systemic problems it is a completely closed environment and it is with no said by the Phil Gibson, CEO of Finger Lakes. But it is a different species, so it is not in direct conflict with Atlantic salmon. Plus, it is at a price premium because RAS technology is more expensive than raising fish in the ocean.

Steelhead trout is another alternative to the Atlantic salmon, at Hudson Valley Fisheries in Hudson, N.Y a salmonid that is flopping around in an RAS system of 54 tanks and 3.5 million gallons of water in a 160,000-square-foot building. In about 16 months they grow from egg to harvest size to restaurants and retail distributors on the U.S. East Coast the company is aiming to peak out at shipping more than 2.2 million pounds of trout each year.

For now, COVID-19 wrecked the company's substantial restaurant business, but to supplement a retail store trade that survived the pandemic Hudson Valley quickly pivoted to e-commerce and local farmers markets. The company also began playing up the advantages of a closed system.

When people think wild-caught is a healthier product, when they are being challenged out, and it is important to help people understand that wild fish carry pollutants such as heavy metals and microplastics, says Hudson Valley Chief Executive Officer John Ng.

To farmed salmon is Australia another company proffering is an alternative, which raises barramundi. In the Vietnam it starts in land-based ponds; then the juvenile fish are transferred to ocean nets. Globally Barramundi is a mild whitefish, but Australis Founder Josh Goldman calls the species "in effect, a tropical salmon," and he says barramundi is the fastest-growing aquacultural species.

In the ponds shrimp is another major aquaculture category grown. For example, by Thai Union North America, chicken of the sea is being owned, in saltwater ponds farms white-legged shrimp which is located close to oceans in countries throughout Asia and Latin America, and freezes them. By the Roxanne Nanninga said that the largest single importer of farmed shrimp who is the director of sustainability for the El Segundo, Calif.–based company.

As the industry ponds were the main sites which took off in the 1970s, in part backed by mighty institutions such as the Rockefeller Foundation. Aqua culturists began to transfer knowledge from agriculture to farming in water.

In ponds one began to see aerators that allow people to control oxygen levels and enabled you to keep big numbers of aquatic organisms alive for longer periods of time, so one could create a farm, says Frank Asche, who is a professor in fisheries and aquatic sciences at the University of Florida.

Moreover, to improve pond environments a relatively new technology known as biofloc has continued. By balancing the carbon and nitrogen in the pond system Biofloc enhances the water quality, with the added value of producing proteinaceous feed in situ.

For the fish farming open-net fencing of areas of oceans also became popular over the last few decades. Until recently ability to fence in the water was very limited, because it is hard to do, Asche says. Yet it is growing because digging a pond is very expensive, and it becomes much more expensive to do than fencing. It is also hard to see what is going on in that body of water.

From the Agri-science in both ponds and the ocean, as soon as one had the ability to create these farms, and control over creatures, and one could start using the whole knowledge system from, including breeding, feeding, and disease management, as said by Asche. As the modern chicken barn, the holy grail was to become as sophisticated.

8.3.4. Feeding Farmed Fish

It has been said that the biggest challenge to aquaculture in any setting is optimizing inputs and outputs, just like with the modern chicken barn. On the availability of fish meal made out of small varieties such as anchovies, so for one thing aqua culturists are facing limits, for large species such as salmon and tuna, mimicking nature which traditionally have been the biggest source of nutrients.

There is a senior research scientist for the Archer Daniels Midland (ADM) John Bowzer who says that fisheries for anchovies have become stagnated because governments have put capacity limits on them. That has to be replaced with other quality ingredients that fit into the diets of the target species.

Right now, into the aquaculture 70% of the world's fish oil supplies go back. On the terrestrial feed crops there is limits on aquatic sources have meant modern aquaculture and increasingly competes for resources with livestock, the energy industry, and direct human consumption. In the innovative sources of supply that pinch has led to fast-rising, including bacteria-based proteins, insect meal, and oil from microalgae.

This the place where the Chicago-based ADM comes in fermenting algae in large tanks like those used in microbrew and ethanol production, and providing algal oil for aquaculture feed. There is another oil provider Netherlands-based Corbion, which says its algal variety contains three times the level of DHA found in fish oil.

As the wild fish people are recreating some of the same nutrient's profiles that salmon normally would eat in the wild, and a replacement for the fish oil and fish meal derived from fish that salmon eat in aquaculture, says Jill Kauffman Johnson, head of global market development for algae ingredients for Corbion. It is a more sustainable feed scenario.

At the same time, how the fish are being fed it does not matter, salmon farming's effects on the ocean environment can be significant, and that has prompted national governments to boost regulatory control. The feed that was not eaten by fish went to the bottom, and the fish below them were eating feces of salmon and sea bass, in the early days if industry.

For the salmon farms now is a lot like the challenges in farming, as Asche says where most developed countries are limiting the number of pigs or cows a farmer can have and how much manure they can get rid of. An environmental impact an ocean pen makes so moderate that they do not have any system-changing impacts. In the areas as long as, one situates them in the areas as one not involving in the overloading system.

In Norway or Chile or the United Kingdom as one produces the salmon says Stein and one is not allowed to just keep throwing more and more net pens into the sea. In the licensed areas there are restrictions on the volume of salmon because of the pollution that comes from these farms.

The salmon industry at least has evolved a learned how to control that problem so that it can have the lowest impact possible, as it is being argued by the Ryan.

Over the environmental sustainability these arguments like are shaping much of the future of aquaculture in the United States and elsewhere. To the land-based agriculture it is an important alternative and by the global population growth which is being is stressed and also by the rising concerns about climate change it is being restricted, and into the service by pressing more marginal lands limited practically.

With the sufficient product and technologies if one can come up that can grow food in the world's waters, and for deforestation one can relieve the pressure on land. This statement is said by Asche. And on the other hand, by people who view harnessing the waters for anything as adding environmental stress there are opposing views. As the new way of infringing, they view aquaculture on the system. But in total, one is doing the favor on the environment and if one is eating a chunk of salmon instead of a steak or a pork chop.

By the climate change aquaculture also is having to adjust to how global waters may be affected. It has been said that spawning grounds will be more deteriorated or will have to shift, as the oceans get warmer. And in the same way aqua culturists will not be able to produce or harvest.

As well as the economic activities, there are such environmental pressures which are generating more innovations in aquaculture. To raise the species rising water temperatures and changes in ocean acidity are prompting more aqua culturists such as blue mussels, and to become resistant to environmental changes which can alter patterns in their genes. In Japan and other countries sea urchins are another type of aqua-farmed species which are resistant to the climate vagaries. For the aquaculture there some companies and the government who are doing research on new opportunities which will result in the reduction of the cost and eve reduce the environmental loads. For instance, for the wild harvesting there is a need to reduce the need of some of the most popular varieties of fish.

From the bluefin there some of the Japanese scientists are transplanting stem cells into smaller, and there are some of the related species that can be raised more quickly via aquaculture and cut the demand for wild-harvested bluefin.

On the species they have been focusing that sexually mature more quickly and have been cultured before, such as mackerel and yellowfin tuna, and can be grown in outdoor pens, says Tawny Scanlan, an FTW Ventures associate who is affiliated with the vaunted aquaculture program at the University of California, Davis. In just five to six years versus 10 to 12 years for bluefin yellowfin is mature.

8.3.5. Serving Up Seaweed

On the increasing aquaculture there is the emphasis of kelp and other sea plants, just like the plants on the land which are being considered more environment friendly than the creatures. It has been seen that the seaweed is low in fat and rich in proteins, carbohydrates, minerals, vitamins, and essential micronutrients. The carbon dioxide is being captured and stored by it and the scourge of climate stress.

This is the reason for the coalition of the groups and also led by the Lloyd's Register Foundation and for a greatly expanded seaweed industry playing a globally significant role in food security this year issued Seaweed Revolution, a "manifesto" proposing a vision for this and along with this also for the climate change mitigation and to support the marine ecosystem, as well as contributing to job creation and poverty alleviation.

In the Asian coastal waters at this point of time nearly all seaweed is cultivated. But by approaching it from the demand end there some of the entrepreneurs are trying to expand seaweed aquaculture. It has been said that except in sushi and some dried snacks western consumers mostly have shunned seaweed, and that is mostly from Asian seaweed varieties called nori. So, in the San Francisco Patrick Schnettler started a company in called 12 Tides that produces organic puffed snacks in flavors such as sea salt and chili, made from kelp seaweeds that are cultivated in North America.

At a pretty extraordinary rate and can grow super-fast kelp absorbs the carbon, more than a foot a day during growing season. From the climate change this model also can mitigate some of the impacts of acidification at least at a local level, which has the implications for the biodiversity.

In the industry cellular aquaculture there is another new frontier which is being cultured seafood, as for aquaculture as well as wild harvesting though it goes to the extreme of negating the need. For instance, from the popular fish it starts with the tissue such as yellowtail and mahi and teases them into pieces of fish analog with the aim of replicating the nutrition and sensory experience of real seafood.

It has been said that to any of the significant degree there is a way which is being considered as too early to tell whether such engineered foods ever will complement wild-caught and aquaculturally raised seafood. Surveys have been conducted that suggests that maybe one or two out of three people will try cultured seafoods, but will they make it part of their diet. It will show up in high-end restaurants that want to take a gamble and show them off. Within two or three years. And mass production is the next stage and no one is sure that it is viable or not.

Between the current demand for seafood and the anticipated needs of the future it will be up to aquaculture to keep trying to close the gap between in the meantime. In terms of health there is a need to improve awareness, as there is no difference between wild and farmed salmon and in the environment, there will be no difference. It has been said that aquaculture is a good alternative as well if anyone cannot find good wild-caught.

The challenge of confronting the suspicions of many consumers about farmed versus wild-caught fish is being included by the further growth of aquaculture, especially in an era when the production narratives behind all foods have become so important to so many. It has been said that the environmental benefits of farmed versus caught seafood could help in that regard. About the sustainability the aquaculture space right is there and a lot of that is being driven by the customer base.

8.4. INTEGRATED MARICULTURE: ITS ROLE IN FUTURE AQUACULTURE DEVELOPMENT

It has been said that within 30 years it seems feasible that the sector will be able to meet the future challenge and double its production looking at the present growth of aquaculture production. Therefore, this expansion will take place is not the interesting question but what the resulting environmental and socioeconomic consequences will be and how it will be achieved.

From the fresh water fish aquaculture there is the increased inland production, which is known as the traditionally an integral part of agriculture production, and a large part of the expansion is also anticipated to take place along the coasts of both developing and developed countries.

The production of more extensive species like bivalves and seaweeds will increase but fish will be the main farmed aquaculture species. Especially in developing countries there are coastal areas who already experience high pressure from, e.g., urbanization, and industrialization that has resulted in both terrestrial and marine ecosystem degradation.

By the culture system characteristics, the potential impacts from any aquaculture development will mainly be determined. Generally, as compared to others some aquaculture systems can be identified as being more sustainable.

Though, from the environmental and socio-economic perspectives there are the needs to be considered in each case of aquaculture development, local constraints. To meet some important aspects of sustainability there are more environment friendly species which may fail under some of the circumstances.

It has been said that increased aquaculture production should come from species to meet the future protein demand and for the human consumption not dependent upon high-food-grade raw materials, i.e., fishmeal, and fish oil produced from fish species suitable.

In order to utilize the vegetarian diets and on extractive species required a complete focus on the species. In the production of fish species, the development of alternative protein feed sources will also play a role today depend on fishmeal-based feeds for a review on feed development.

In the human activities the rapid-scale increase now has been seen and in the coastal areas it puts further pressure on the already impoverished functions of coastal ecosystems. Though, regarding the discharge of nutrient wastes from various sources there are some legislative guidelines, standards, and controls which are starting to become more stringent in many countries.

Through the integrated concepts there is the development of integrated mariculture, i.e., bioremediation with a capacity to improve the quality of the discharged water, may facilitate the aquaculture industry to avoid non-compliance and gain both direct and indirect benefits from improving water quality and coastal ecosystem health. In many recent studies the finding states that for adopting integrated mariculture systems both land-based and open marine culture systems show the potential. In the year 2003 European Aquaculture Society meeting "Beyond Monoculture," held in Trondheim, Norway, which conclude that one to accumulate enough data to support the biological demonstration of the concept of integrated aquaculture.

In addition to it, there is no universal system that exist and the choice of technology and species is different for the different regions and different socio-economic conditions and that now need to demonstrate its applicability at commercial scales, including analyses of both biological and economic performance. It has been said that today only a few commercial integrated mariculture systems exist, despite of some 20 years of research.

At the time of when extrapolating the results most recent marine integrated systems have been experimental and small scale, which implies difficulties to commercial scales. Though, at the larger scale there are the integrated systems that are operating and being commercial.

In China the best can be found where suspended multi-species aquaculture operates at scales of whole bays. For instance, scallops are cultivated together with kelp, abalone, and fish in Sungo Bay, east of the Shandong Peninsula and the cultures extending 8 km offshore.

With the salmon and mussels another integrated open-water culture of kelp operates in the Bay of Fundy, Canada. In the south Africa and Israel, the other examples of integrated mariculture systems performed at larger scales on land. In the land-based abalones are cultured and flow-through tank systems and seaweeds are cultured in the wastewaters from the abalones.

It has been said that from co-cultured crops and reduction of nutrient release to the environment the general benefits from integrated practices are additional income. For the recirculation of waters there are the other benefits such as facilitation, which also could result in reduced pumping costs and increased water temperature.

For a limited period, it has been said that the ability to operate in recirculation mode may be important in cases where intake water may need to be shut. An economic value can be generated by the cultivated additional crop, i.e., the extractive organism and can be used as input to co-cultured species.

With the abalone's seaweeds can be cultivated and are used as abalone feeds, and their quality has been proven to increase when cultured in abalone wastewater. To a large extent on wild-harvested kelp for the species like abalones that in some countries and in the future expansion may depend on increased cultivation of seaweeds when wild kelp resources become exhausted.

There are some of the additional arguments which are based on integrated mariculture and include possible social benefits and diversified production. Additional cost may be generated by the practice of integrated farming that may impact negatively on overall farm economy. Some of the additional investments, maintenance, and the need for increased and complementary skills these costs include.

For the negative interactions between the species there might also be an increased risk. When practicing integration at large scales these potential costs and interactions may only be revealed. As many beneficial interactions probably are limited by scale, it is also important. For their practice are weak the fact that few integrated mariculture systems exist could indicate that the incentives.

Business models do not consider and it is important to acknowledge that present aquaculture and recognize the economic value of the bio mitigation services provided by biofilters, in mariculture systems as there is no cost associated with aquaculture discharge/effluent.

To clearly recognize the benefits of the extractive components in a culture system the regulatory and financial incentive may therefore be required. The industry to incest in the integrated approaches there is a better estimate of the overall cost/benefits to nature and society of aquaculture waste and its mitigation would create powerful financial and regulatory incentives to governments.

8.5. CONCLUSION

Overall, it has been that there is increase in the income and wealth, healthier diets, and population growth are all leading to increased consumption of seafood. From the oceans the inability to extract more seafood and will drive the need for increases in aquaculture production. In helping meet growing demand aquaculture, or farmed seafood, has great potential while protecting wild fish stocks, and at the same offering a healthy and eco-efficient protein source.

And if there is a comparison made with the other forms of protein production, it has been seen that aquaculture is intrinsically more sustainable, in regard to both environmental impact and resource use. Though, to meet growing demands for protein and considerable changes aquaculture production must grow substantially, at speed and at scale, are needed to ensure the environmental impact of this needed growth is sustainable.

REFERENCES

1. Alessandro, L., Micheal, J. P., Richard, A. J., & Koji, Y., (2008). *The Future of Mariculture: A Regional Approach for Responsible Development in the Asia-Pacific Region*. [eBook] Guangzhou, China: Food and Agriculture Organization of the United Nations. Available at: http://www.fao.org/3/i0202e/i0202e.pdf (accessed 10 September 2021).
2. Alex, B., (2018). *Mariculture: Farming the Ocean to Feed a Future Population of 10 Billion Humans*. [online] Medium. Available at: https://medium.com/predict/mariculture-farming-the-ocean-to-feed-a-future-population-of-10-billion-humans-deb975abaa3d (accessed 10 September 2021).
3. Costello, C., Cao, L., Gelcich, S., Cisneros-Mata, M., Free, C., Froehlich, H., et al., (2020). The future of food from the sea. *Nature*, [online] *588*(7836), pp. 95–100. Available at: https://www.nature.com/articles/s41586-020-2616-y (accessed 10 September 2021).
4. Dale, B., (2020). *Sea Change: New Horizons in Aquaculture*. [online] Institute of Food Technologists. Available at: https://www.ift.org/news-and-publications/food-technology-magazine/issues/2020/november/features/sea-change-new-horizons-in-aquaculture?gclid=CjwKCAjwtdeFBhBAEiwAKOIy5x9HyHSGhA2-djWfk6m1R4E1lUGkFfY4qxkJucr7STNDXx1XILI68RoCMXUQAvD_BwE (accessed 10 September 2021).
5. Frerck, R., (n.d.). *The Future of Aquaculture, Two Sides of the Story*. [online] Blueocean.net. Available at: https://blueocean.net/future-aquaculture-two-sides-story/ (accessed 10 September 2021).
6. Global Salmon Initiative, (2021). *The Future of Aquaculture*. [online] Global Salmon Initiative. Available at: https://globalsalmoninitiative.org/en/our-work/the-future-of-aquaculture/ (accessed 10 September 2021).

INDEX

W

Printed in the United States
by Baker & Taylor Publisher Services